Notes:

Fabrication

of

Metal Hull Integrals

By way of Bend Deduction
And
Bend Allowance

D L. Schaffer

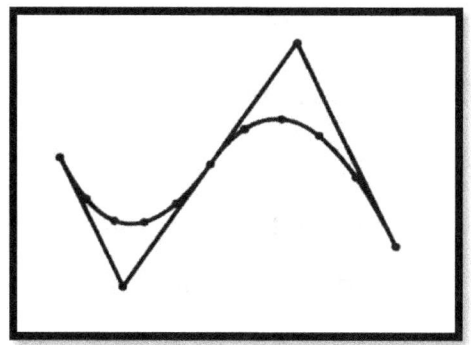

Part I - Fundamentals:

Part II - Preliminary Calculations:

Part III – Formulas:

Part IV – Deck Hatches:

Part V – Port Lights:

Part VI – Summary:

Part I
Fundamentals

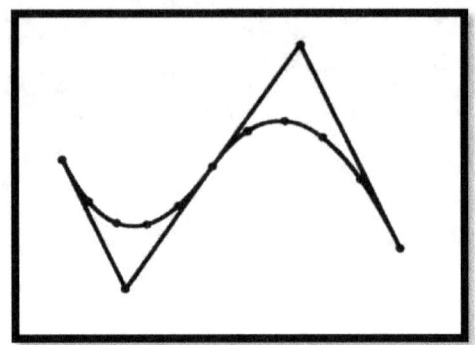

Bend Angles:

Sheetmetal can be bent to any angle from zero to 180 degrees. The angle of the bend is always measured from the flat. In metal fabrication obtuse angles, *Figure 1-01*, have a bend angle from zero to something shy of 90 degrees.

Figure 1-01

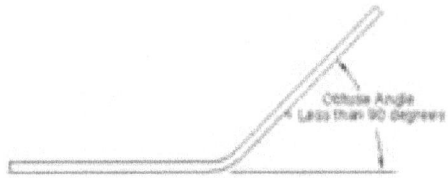

A 'Right Triangle', *Figure 1-02*, is where one (1) angle of the Right Triangle is exactly 90 degrees.

Figure 1-02

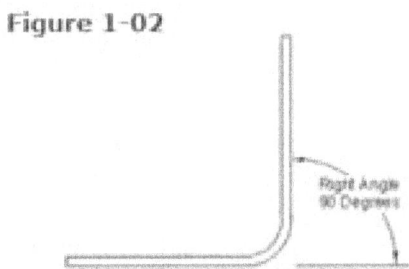

Acute angles, *Figure 1-03*, are bend angles something over 90 degrees to 180 degrees.

Figure 1-03

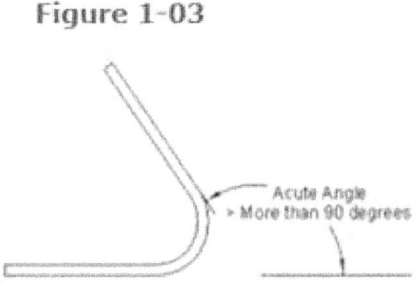

Bend Terminology:

Figure 1- 04, 1-05, and 1-06, illustrates the fundamental parts of a formed angle in sheet material. No matter the bend angle (obtuse - right - acute) all have the same characteristics listed below:

- **Material Thickness** – Thickness of material.
- **Inside Radius** – The Inside Radius of the bend.
- **Start of the radius** - The dividing point between the radius section of the bend and the straight.
- **Straight -** The distance from the start of the radius to the edge of the part or the distance between the starts of the radii.
- **Bend Allowance -** The calculated distance in the radius section of the part.
- **Outside Apex -** The extension and crossing point of each side of the bend on the outside of the material.
- **AIR -** The distance between the Outside Apex and the Inside Radius of the bend.
- **Inside Material Apex -** The extension and crossing point of each side of the bend on the inside of the material.

Figure 1-04

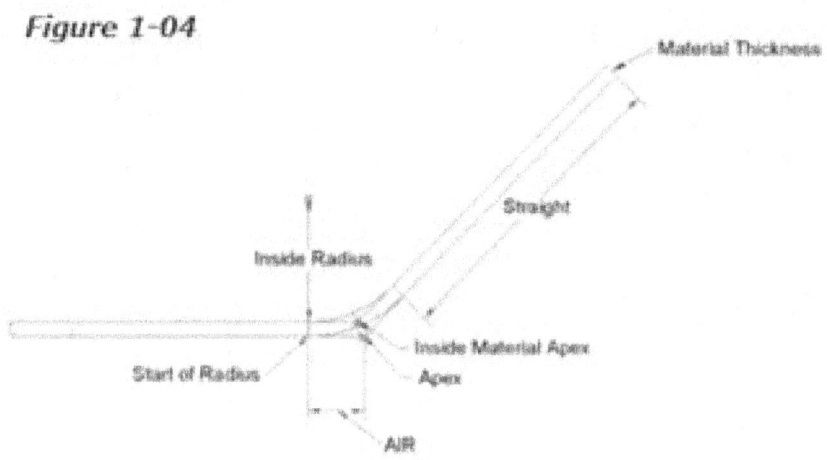

Figure 1-05

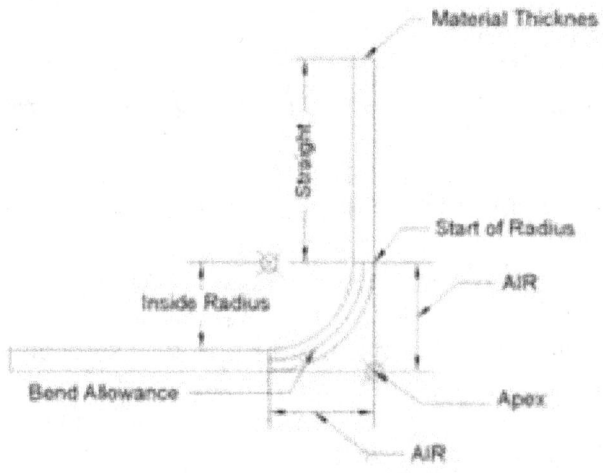

Figure 1-06

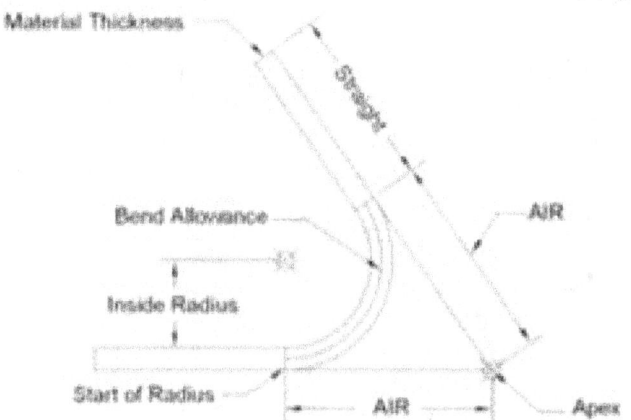

Bend Dimensioning:

There are a number of ways formed sheet-metal parts are dimensioned in a drawing:

- ☐ Dimensioning to the 'Outside Apex' or between them.
- ☐ Dimensioning to the 'Start of the Radius' or between them.
- ☐ Dimensioning 'Tangent to the Radius' or between them.
- ☐ Dimensioning to 'Inside material Apex' or between them.
- ☐ Or any combinations of the above.

Obtuse Angle Bends:
Obtuse Angle bends can be dimensioned from:

- ☐ The outside apex to the end of the part as shown in, **Figure 1- 07**, or between outside apexes.
- ☐ From the start of the radius, **Figure 1-07**, to the end of the part or between the start of the radii.

- ☐ From the inside material apex, **Figure 1-08**, to the end of the part or between the inside material apexes.

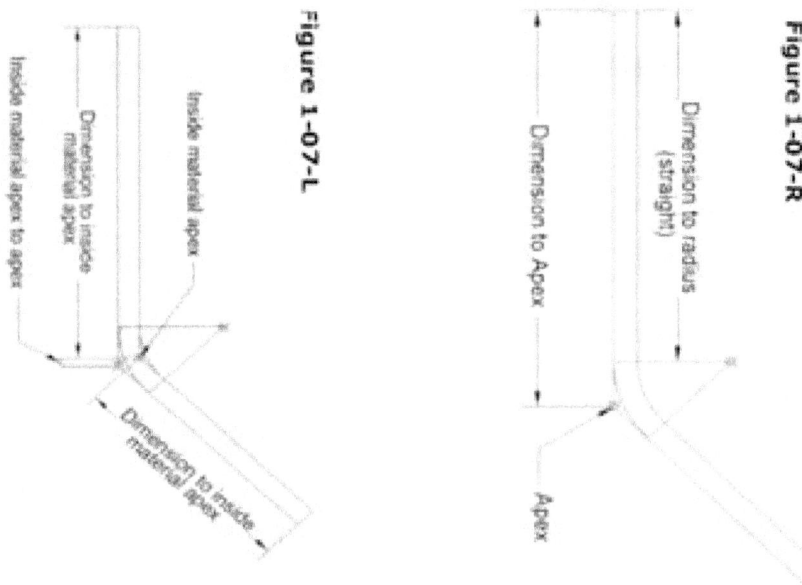

Right Angle Bends:

Right Angle bends are dimensioned from:

☐ The outside apex to the end of the part as shown in, **Figure 1- 09**, or between outside apexes.
☐ From the start of the radius, **Figure 1-09**, to the end of the part or between the start of the radii.
☐ From the inside material apex, **Figure 1-10**, to the end of the part or between the inside material apexes.

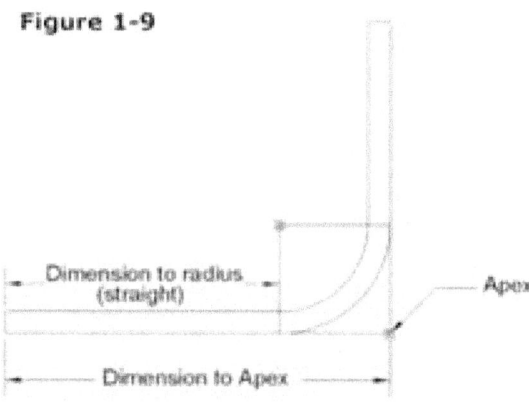

Figure 1-9

Dimension to radius (straight)

Apex

Dimension to Apex

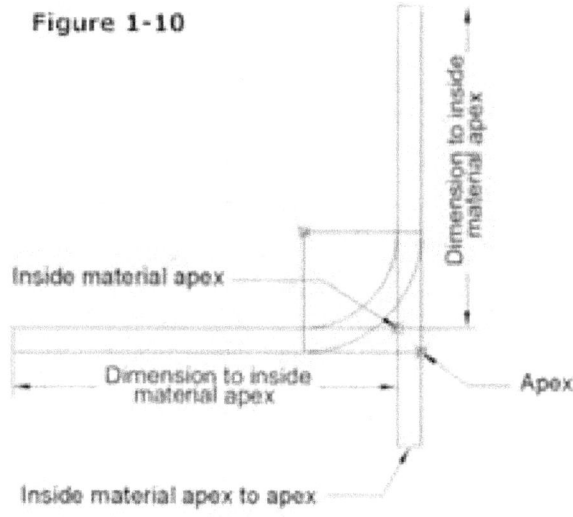

Figure 1-10

Dimension to inside material apex

Inside material apex

Apex

Dimension to inside material apex

Inside material apex to apex

Acute Angle Bends:

Acute Angle bends are dimensioned from:

☐ From the outside apex to the end of the part as shown in, **Figure 1-11**, or between outside apexes.

☐ From the start of the radius, **Figure 1-11**, to the end of the part or between the start of the radii.

☐ Tangent to the radius to the end of the part or between the tangents of the radii.

☐ From the inside material, **Figure 1-12**, apex to the end of the part or between the inside material apexes.

Figure 1 - 11

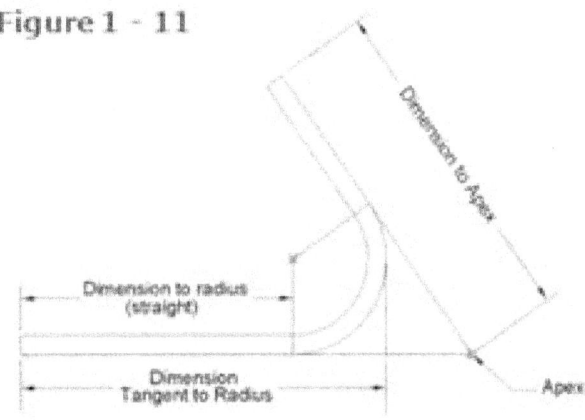

Figure 1 - 12

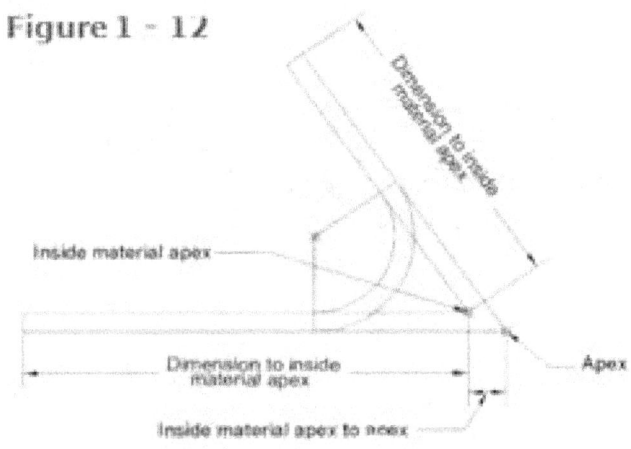

Press-Brakes Tooling:

Press-brakes, *Figure 1-13*, are open ended Sheetmetal forming machines and can accommodate a large variety of tooling to form custom integrals such as Hatch's and Window for your Build.

Figure 1-13

The list of tooling in *Figure 1-14* are common to Metal Fabrication Shops and will be used to bend the example fabrications detailed in this book. The top tool is known as the Punch and the bottom tool is known as the Die.

□ Type 'A' - Straight punch and die sets come in varying thickness and used when tight Inside Bend radii are required.

□ Type 'B' - Offset punch and die sets also come in varying thickness. They are used when the distance between bends become a factor and when tight 'Inside bend radii are required.

□ Type 'C' - Radius top punches are used when large inside radius bends are required.

• Type 'D' - Heming punch and die sets are used when bending angles over 90 degrees.

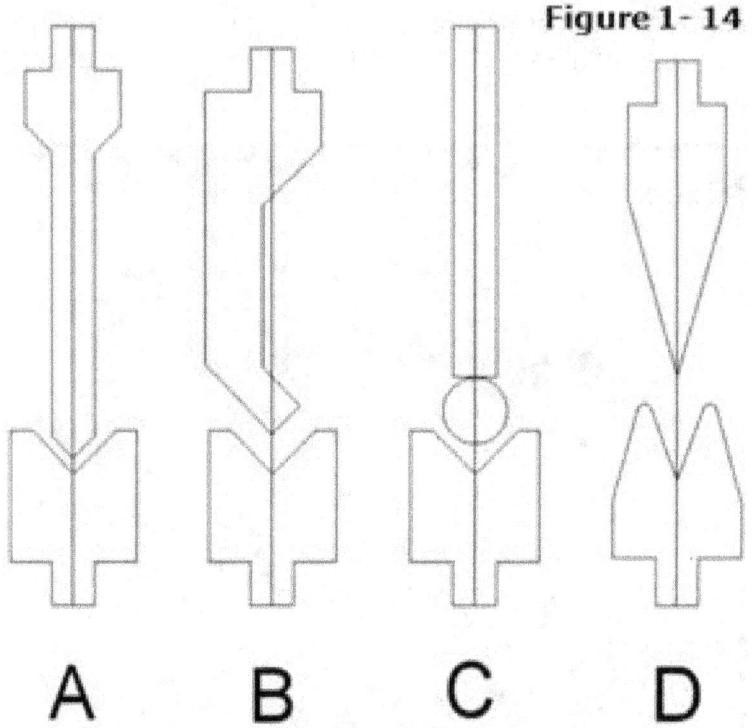

Figure 1- 14

A   B   C   D

Notes:

Part II
Preliminary Calculations

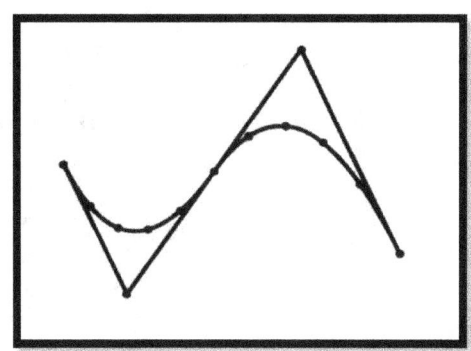

Inside Radius Calculations:

The 'Inside radius' for a 90-degree bend using a Vee top-punch configurations, in the Air Bending Mode, **Figure 2-01** (left), is determined by a percentage (15.6%) of the bottom dies opening when forming steel. The formula is:

Bottom die width x .156 = Inside Radius of the Bend

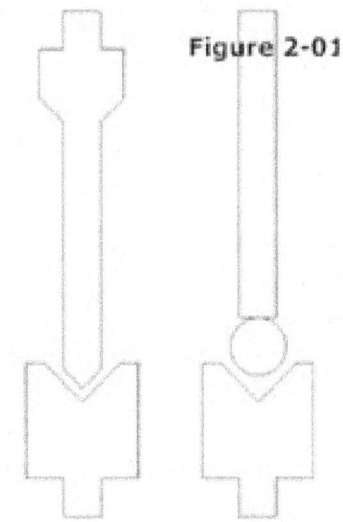

Figure 2-01

When round top punches, **Figure 2-01** (right), are chosen the'Inside Radius' will equal the radius of that punch, therefore the opening of the Vee bottom die will be at least the diameter of the top punch plus two material thickness.

For comparison, the 'Inside Radii' for different bottom die opening ranging from 1.250" to 0.500" is listed below.

Bottom die opening x .156 (15.6%) = Inside Radius

$$1.250" \times .156 = .195"$$
$$1.000" \times .156 = .156"$$
$$0.875" \times .156 = .136"$$
$$0.750" \times .156 = .117"$$
$$0.625" \times .156 = .097"$$
$$0.500" \times .156 = .078"$$

It is easily seen that the bigger the opening of the bottom Vee die, the larger the Inside Radius will be. The question is: Which vee bottom die opening should be used? The answer is found in the Bending Mode.

Bending Modes:

There are two Bend Modes: Air Bending and Bottom Bending. The point of demarcation between the two is somewhat elusive, and is one subjects of the book.

☐ When in the Air Bending Mode, *Figure 2-02-Left*, there is always a three-point contact between the top punch and the bottom die. The top punch and bottom die never make full contact, resulting in the least amount of tonnage to produce a 90-degree bend.

☐ In the Bottom Bending Mode, *Figure 2-02-Right*, there is full contact between the top punch and bottom die when forming 90-degree bend angles. The material conforms to the die and inside radius of the tooling. Press-brake pressures require to Bottom Bend a 90-degree angle are greatly increased many times over Air Bending. Bottom Bending is primarily use when the inside radii of the bend is small.

It is established practice to work within the **Air bending** range, where there is a three-point contact between the top punch and bottom die as opposed to full contact between the top punch and bottom die when forming a 90-degree bend. However, this should not prevent us from working in the Bottom Bend mode when required.

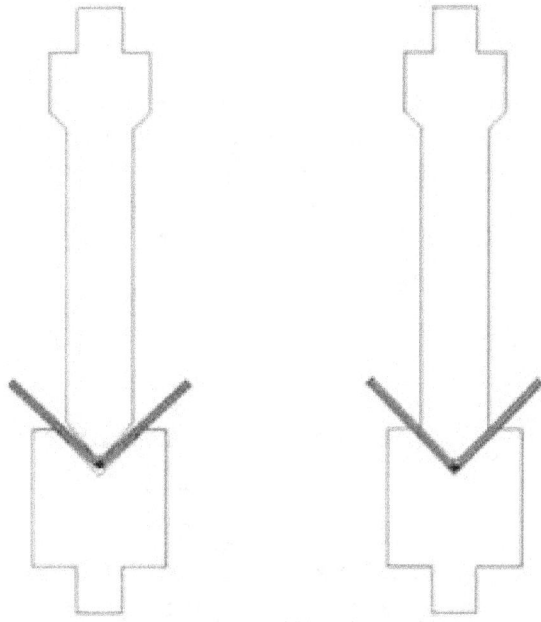

Figure 2 - 02

The Bending Mode is determined by dividing the Inside Radius by the material thickness. Any percentage of thickness at or below 81% is considered **Bottom Bending**.

The following table determines the percentage for 11ga - .119" thick material.

Inside Radius / Material Thickness = % of Material Thickness

.195" / .119" = 1.63 - 163%
.156" / .119" = 1.31 – 131%
.136" / .119" = 1.14 – 114%
.117" / .119" = 0.98 – 98%
.097" / .119" = 0.81 – 81%
.078" / .119" = 0.65 – 65%

From the above results we can conclude that a die opening of 0.750" would place the bend solidly in the Air Bending mode. Choosing a die opening of 0.625" would be on the cusp between Air and Bottom Bending.

There also is a 'Rule of thumb' method for choosing the bottom vee die opening for a given material thickness that places a bend solidly within the Air Bending mode.

The rule of the thumb states that the bottom die opening should be eight (8) times the material thickness to be solidly in the 'Air bending' mode. Seven (7) times the thickness would be leaving the 'Air bending' mode, six (6) times the thickness would be entering the 'Bottom bending' mode, and Five (5) times the thickness would place us solidly in the 'Bottom bending' mode.

Using a material thickness of .119" or 11 gauge the "Bottom Die Opening' is calculated as below:

Material Thickness x varies = Vee bottom die opening

☐ .119 x 8 = 0.952"
☐ .119 x 7 = 0.833"
☐ .119 x 6 = 0.714"
☐ .119 x 5 = 0.595"

Round the above results to the nearest standard bottom vee die opening to:

☐ .119 x 8 = 1.000"
☐ .119 x 7 = 0.875"
☐ .119 x 6 = 0.750"
☐ .119 x 5 = 0.625"

The proceeding calculations are relatively simple, however as we proceed the calculations will be more complex. As the calculations become complex, they become human error prone. To circumvent this Excel Spreadsheets have been created for all the calculations in this book.

These Excel spreadsheets can be instantly download to any email address directly from metalsailboats.com.

Figure 2-03, is a screen-shot of the Spread-sheet use to calculate and all the preceding formula, while revealing the relationship between the Material thickness, Bottom die opening, Inside radius, their relationship to Air and Bottom bending modes.

The convention use of input into the Spreadsheets are:

- The unboxed cells are for User Input.
- Boxed cells and all other cells in the spread-sheet are locked to prevent accidental changes to the Spreadsheets formulas.
- Placing your mouse over numeric open and boxed cell will reveal the purpose of that cell.

Open the 'Inside Radius' Spreadsheet shown in *Figure 2-03.*

- Enter a material thickness and a bottom die opening. Enter a bottom die opening that you think would be appropriate.
- The following four (4) closed cells will calculate the die opening vs the material thickness for several Die openings to assist you to fine tune your choice of Die openings.
- The next closed cell will calculate the 'Inside Radius' for your chosen 'Bottom Die' opening.
- The next closed cell will be Bottom Die vs Material thickness for the bottom die opening you chose.
- The final closed cell is the percentage indicator.

Selecting bottom die opening seven (7) to five (5) times the thickness may require 'Test Bending' depending on your working tolerance.

- Working tolerances of plus or minus 0.005" would require a 'Test Bend' calculation, discussed later in the Book.
- Considering the tolerance in our work at plus or minus 0.031", entering the Bottom Bending Mode is not going to taint the outcome.
- Bottom Bending is common practice in Metal Fabrication under many circumstances, just be aware on the Bending Mode.

INSIDE RADIUS CALCULATIONS
FOR
AIR - BOTTOM BENDING RANGE

Figure 2-03

MATERIAL THICKNESS: | 0.119 |
BOTTOM DIE OPENING: | 1.000 |

BOTTOM DIE OPENING AT 8 X MATERIAL THICKNESS: | 0.952 |
BOTTOM DIE OPENING AT 7 X MATERIAL THICKNESS: | 0.833 |
BOTTOM DIE OPENING AT 6 X MATERIAL THICKNESS: | 0.714 |
BOTTOM DIE OPENING AT 5 X MATERIAL THICKNESS: | 0.595 |

AIR TO BOTTOM BENDING RANGES

INSIDE RADIUS: | 0.156 |

BOTTOM DIE OPENING VS MATERIAL THICKNESS: | 8.4 |
BOTTOM BENDING PERCENTAGE INDICATOR: | 131% |

As a comparison, *Figure 2-04,* I have entered a:

- Bottom Die opening of 0.625".
- The 'Inside Radius' is now 0.098".
- The bottom die opening has also been verified at 5.3 time the material thickness.
- The percentage indicator at 82%.
- Clearly the Bottom Bending Mode.

Figure 2-04

INSIDE RADIUS CALCULATIONS
FOR
AIR - BOTTOM BENDING RANGE

MATERIAL THICKNESS: 0.119
BOTTOM DIE OPENING: 0.625

BOTTOM DIE OPENING AT 8 X MATERIAL THICKNESS: 0.952
BOTTOM DIE OPENING AT 7 X MATERIAL THICKNESS: 0.833
BOTTOM DIE OPENING AT 6 X MATERIAL THICKNESS: 0.714
BOTTOM DIE OPENING AT 5 X MATERIAL THICKNESS: 0.595

AIR TO BOTTOM BENDING RANGES

INSIDE RADIUS: 0.098

BOTTOM DIE OPENING VS MATERIAL THICKNESS: 5.3
BOTTOM BENDING PERCENTAGE INDICATOR: 82%

Tonnage Calculations:

To recap we find that for 11 ga - .119" thick material using a bottom vee die opening of 1.000" places us squarely in the 'Air Bending' mode with an 'Inside radius' of .156".

Entering this information into the 'Tonnage chart', **Figure 2-05**, we will cross the die opening of 1.000" along the top edge with the material thickness of 11gauge - .120" along the side edge. The crossing results in a tonnage of 7.4.

This means that it will take 7.4 tons of pressure to form a 90 degree 'Air Bend' in steel 12" long. If for example, our part was 8 feet long we would multiply 7.4 tons x 8 resulting in 59.2 tons of pressure from the press-brake. On this chart the inside bend radius and minimum flange dimension are provided along the bottom of the chart.

Figure 2-05 — Tonnage Chart (THICKNESS across top / WIDTH OF FEMALE DIE down side)

Min Flange	Bend Radius	1/2 .500	7/16 .437	3/8 .375	5/16 .312	1/4 .250	7 .188	8 .160	10 .135	11 .129	12 .105	13 .090	14 .075	16 .060	18 .048	20 .036	22 .030	Width of Female Die
3/16	1/32															2.9	(1.9)	1/4
7/32	3/64														4.0	(2.1)	1.4	5/16
1/4	1/16													5.6	(3.0)	1.7	1.0	3/8
5/16	5/64												6.0	(3.7)	2.2	1.3	0.8	1/2
7/16	3/32										10.1	6.6	(4.5)	2.7	1.6	1.0		5/8
1/2	1/8									10.5	7.4	(5.0)	3.4	2.2	1.3			3/4
9/16	9/64								11.3	8.6	(6.0)	4.3	3.0	1.7				7/8
5/8	5/32							13.1	9.6	(7.4)	5.4	3.7	2.5					1
11/16	11/64						16.4	11.9	(8.3)	6.2	4.4	3.3	2.1					1-1/8
3/4	3/16					28.8	14.0	(9.2)	7.0	5.4	4.0	2.9						1-1/4
13/16	13/64				38.0	22.0	(12.0)	6.7	5.6	4.3	3.2							1-1/2
7/8	7/32			41.0	25.0	(16.0)	7.6	6.2	4.1	3.2								2
15/16	15/64		45.2	29.0	(20.0)	11.5	8.8	3.5										2-1/2
1	1/4	47.9	36.0	(24.0)	16.0	9.1	4.5											3
1-1/8	9/32	39.0	(29.0)	19.4	12.5	7.5												3-1/2
1-1/4	5/16	(32.0)	24.0	16.0	10.6	6.2												4

Figure 2-06, is a screen-shot of the spreadsheet use to calculate the press brake tonnage required to form a part. Again, user input is indicated by open cells. All other cells are locked. Placing the mouse over numeric open and boxed cell will reveal comments on that cell.

PRESS BRAKE TONNAGE

Figure 2-06

MATERIAL FACTORS:

STEEL	=	1.000
STAINLESS STEEL	=	1.400
HARD ALUMINUM	=	0.800
SOFT ALUMINUM	=	0.500

MATERIAL THICKNESS:	0.119
BOTTOM DIE OPENING:	1.000
BOTTOM DIE OPENING VS MATERIAL THICKNESS:	8.4

| MATERIAL FACTOR: | 1.000 |
| LENGTH OF PART IN INCHES: | 12.000 |

| TONNAGE PER INCH: | 0.631 |
| TOTAL TONNAGE: | 7.576 |

If, for example we want to bottom bend this part we would change the 'Bottom die opening' to .625". ***Figure 2-07***, shows the results of this die configuration.

It is easily seen that changing to a smaller bottom die opening places the bend in the bottom-bending mode. This is indicated by the result of the 'Bottom die opening vs material thickness' calculation at (5.3).

It can also be seen the total tonnage for the same 12" long fabrication is increased to (12.122) tons when in the bottom bending mode.

Figure 2-07 PRESS BRAKE TONNAGE

MATERIAL FACTORS:	
STEEL	= 1.000
STAINLESS STEEL	= 1.400
HARD ALUMINUM	= 0.800
SOFT ALUMINUM	= 0.500

MATERIAL THICKNESS:	0.119
BOTTOM DIE OPENING:	0.625
BOTTOM DIE OPENING VS MATERIAL THICKNESS:	5.3

MATERIAL FACTOR:	1.000

LENGTH OF PART IN INCHES:	12.000

TONNAGE PER INCH:	1.010
TOTAL TONNAGE:	12.122

Round Punch Calculations:

Radius top punches, **Figure 2-08**, always operate in the 'Air Bending' Mode. The 'Inside Radius' of the formed material is equal to the radius of the top punch. We do however need to calculate the minimum die opening for a particular radius top punch and material thickness.

(Radius of the top punch x 2) + (2 x Material Thickness)

= Minimum Bottom die Opening

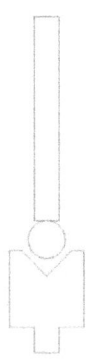

Figure 2-08

Using the spread-sheet, **Figure 2-09**, to find the minimum bottom vee die opening for a 1/2" radius top punch to form .119" thick material is 1.238". We round up to 1.250" as this is the nearest standard die opening.

RADIUS TOP PUNCH
MINIMUM BOTTOM DIE OPENING CALCULATION

RADIUS TOP PUNCH:	0.500
MATERIAL THICKNESS:	0.119
DIAMETER TOP PUNCH:	1.000
MINIMUM VEE BOTTOM DIE OPENING:	1.238

Figure 2-09

Continuing on we can either use the tonnage chart, **Figure 2-05**, by crossing the material thickness with the bottom die opening. Or use the Tonnage spread-sheet, **Figure 2-10**, to calculate press-brake pressure for the part a one foot long.

PRESS BRAKE TONNAGE

MATERIAL FACTORS:

STEEL	=	1.000
STAINLESS STEEL	=	1.400
HARD ALUMINUM	=	0.800
SOFT ALUMINUM	=	0.500

MATERIAL THICKNESS:	0.119
BOTTOM DIE OPENING:	1.250
BOTTOM DIE OPENING VS MATERIAL THICKNESS:	10.5
MATERIAL FACTOR:	1.000
LENGTH OF PART IN INCHES:	12.000
TONNAGE PER INCH:	0.505
TOTAL TONNAGE:	6.061

Figure 2-10

Bump Forming:

Bump Forming, **Figure 1-11**, can be accomplish using types A thru D punch and die sets. Generally speaking, the smoother the bump formed fabrication the better. Type A and B top punches would yield acceptable results. Type D top punches probably would result in more marking of the material. Type C top punches would result in the least surface markings as a result of the press-brake process. Your metal fabricator could advise you to their standard practice. Bump Forming is a common practice and cost effective in the custom metal fabrication industry for 'One Of' fabrications.

Bump Forming is used to form the true round shell plating of Bezier Chine designs as detailed in **'True Round Metal Boat Building'**.

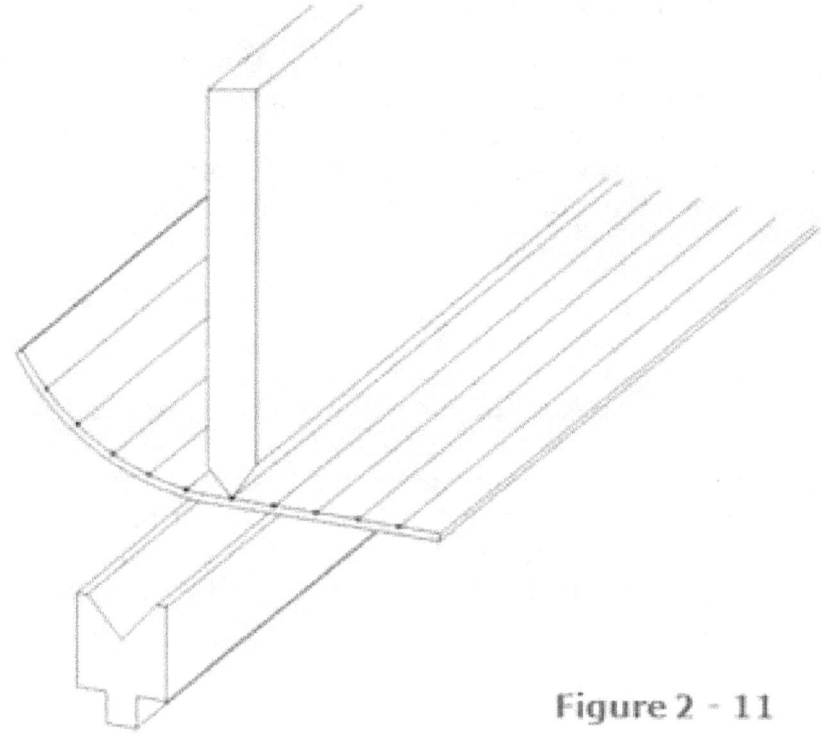

Figure 2 - 11

Tooling - Common Bend Conflicts:

As a 'One-of' steel boat builder we will be forming our fabrication using standard press-brake tooling. There are however tool and bending basic principles that you, the builder, needs to be aware of when choosing the bottom die opening as related to the bend depth.

Figure 2-12, Left, indicates that the material must sit firmly on both side of the bottom die. For example, using a 1.000" open bottom the bend line can be no smaller than half the die opening plus the distance required for the material to sit firmly on the opposite side of bottom die. In this case the bend line can be no less than approximately .562".

Using a smaller bottom die will decrease that distance, but will increase press-brake pressure for the given material thickness. This may on not make a difference. Always keep track of what Bending Mode you will be operating under.

Another common concern is bend geometry that cannot be formed because bends are too close together as in, **Figure 2-12,** Center. The second bend being too small causes the first bend to contact the top punch resulting in deformation of the part. To correct, increase the length of the second bend. This could affect the designed purpose of the part.

Another alternative is shown in, **Figure 2-12**, Right, where an offset top punch is used. It should be noted that straight top punches are rated for more tonnage than the same offset punch. Depending on the material thickness an offset top punch may not be rated for the job.

It may prove wise to visit your fabricator to discuss their available tooling and press-brake specifications before laying out any of your parts. This would allow you to design around their readily available tooling where practicable or chose another metal fabrication shop where your requirement can be had.

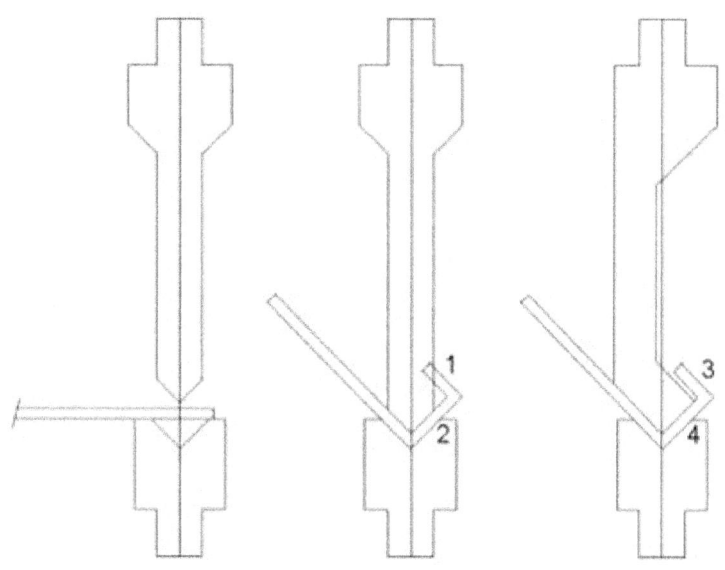

Notes:

Part III
Formulas

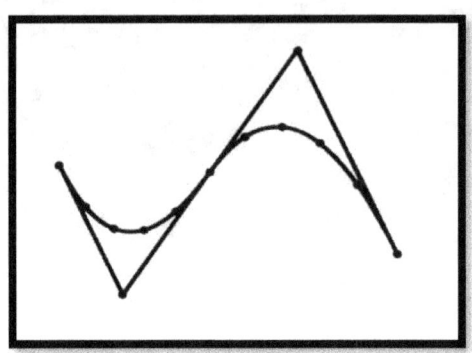

Formula Accuracies:

'Bend Allowance' and 'Bend Deduction' are corresponding formulas used to layout Sheetmetal fabrications onto a single plane for forming. Both formulas are empirical and have been refined to holding tolerances of plus or minus .005" in mild steel. Mild steel is a common name for the steel we will be using throughout hull and deck construction.

Since 'Bend Allowance' and 'Bend Deduction' are capable of calculating cut size that can hold dimensional tolerance to plus or minus .005" in steel construction other material will certainly give results that are within our practicable working tolerance of plus or minus .032".

For those who seek perfection in their calculations there are 'Test Bending' methods that use the principles of 'Bend Deduction' that will deliver plus or minus .005" tolerance for the material of choice.

Bend Allowance:

Sheetmetal stretches on one side of the material and compresses on the other side of the material as it is being formed. The smaller the radius of the bend the more it stretches and compresses.

'Bend Allowance' is the dimensional adjustment that compensates for the stretching or compression of the material while it is being formed. The formula that calculates this adjustment is empirical. The constants in the formula were developed by trial and error until a combination was found that resulted in tolerances of plus or minus .005" in steel.

The Formula is:

BA = ((.0078 x MT) + (.01743 x IR)) x A

Where

- **BA** = **Bend Allowance**
- **MT** = **Material Thickness**
- **IR** = **Inside Radius**
- **A** = **Angle of the Bend**
- **.0078** = **Constant**
- **.01743** = **Constant**

To visualize, **Figure 3-01**, illustrates a three-dimensional view of a fabricated part that will be use throughout to demonstrate 'Bend Allowance', 'Bend Deduction' and 'Test Bending' methods.

Figure 3-01

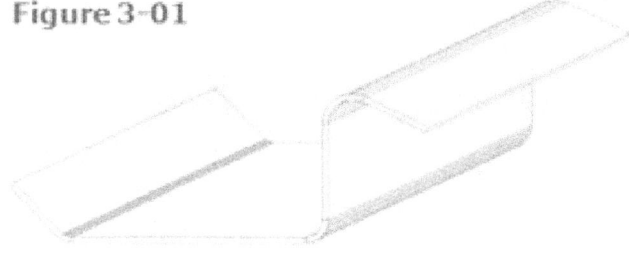

Figure 3-02, is the dimensioned drawing for this part. It could have been drawing by the Designer or it could have been drawn by you the builder.

The drawing has been dimensioned between the ends of the part and the start of a radius, and between radii. Parts dimensioned this way are easiest developed by 'Bend Allowance'.

Note that the part has three different bend angles and three different inside radii for each bend. The formula for 'Bend Allowance' is:

$$((.0078 \times MT) + (.01743 \times IR)) \times A = BA$$

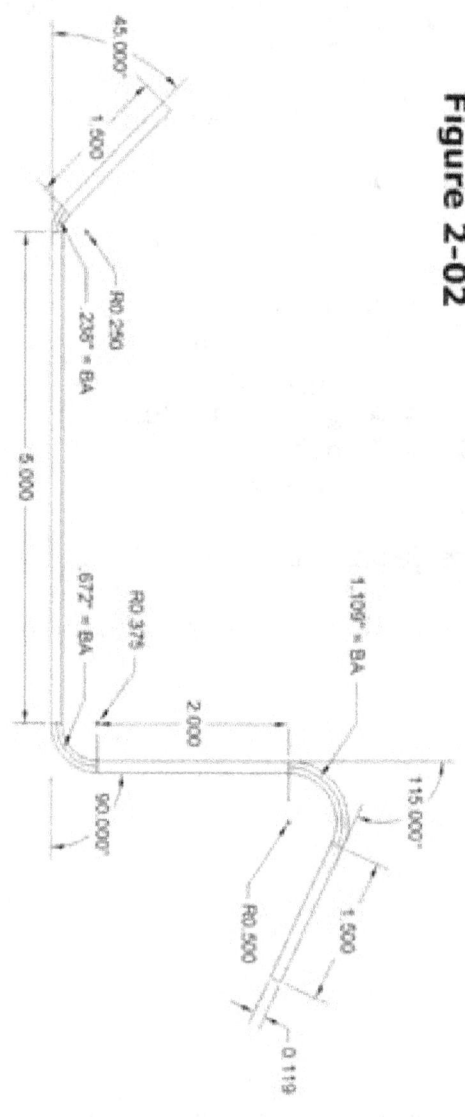

Figure 2-02

Therefore

$$((.0078 \times .119) + (.01743 \times .250")) \times 45 = .238"$$

$$((.0078 \times .119) + (.01743 \times .375")) \times 90 = .672"$$

$$((.0078 \times .119) + (.01743 \times .500")) \times 115 = 1.109"$$

To calculate the cut size: Add the calculated Bend Allowances and the Straight dimension between the radii together. The result is: **12.019"**.

As an alternative to the Bend Allowance formula; use the chart shown in **_Figure 3-04_**.

By crossing the chosen material thickness along the top row of the chart with the chosen 'Inside Radius' along the left side of the chart the "Bend Allowance' for one (1) degree is shown. Multiply this number by the number of degrees in the bend. Let us apply the chart to our part.

.250" IR and .119" MT find .00528 x 45 = .238"

.375" IR and .119" MT find .00747 x 90 = .672"

.500" IR and .119"MT find .00965 x 115 = 1.109

Figure 3-04 BEND ALLOWANCE — STEEL SHEET

GAUGE	24	22	20	18	16	14	12	11	1/8	10	5/32	3/16	1/4	5/16
Material Thickness	.0239	.0299	.0359	.0478	.0598	.0747	.1046	.1196	.125	.1345	.1562	.1875	.250	.3125
Inside Radius	BEND ALLOWANCE FOR 1° ANGLE													
1/32	.00073	.00077	.00082	.00091										
1/16	.00128	.00132	.00138	.00146	.00155	.00167								
3/32	.00182	.00187	.00191	.00200	.00210	.00221	.00245							
1/8	.00237	.00241	.00247	.00255	.00265	.00276	.00299	.00311	.00315					
5/32	.00291	.00295	.00300	.00309	.00319	.00330	.00354	.00365	.00370	.00377	.00394			
3/16	.00345	.00350	.00355	.00364	.00373	.00385	.00408	.00420	.00424	.00432	.00449	.00473		
7/32	.00400	.00404	.00409	.00418	.00428	.00439	.00463	.00474	.00479	.00486	.00503	.00527		
1/4	.00454	.00458	.00463	.00472	.00482	.00493	.00516	.00528	.00533	.00540	.00557	.00581	.00630	
9/32	.00509	.00513	.00518	.00527	.00537	.00548	.00572	.00583	.00588	.00595	.00612	.00636	.00685	
5/16	.00563	.00568	.00573	.00582	.00591	.00603	.00626	.00638	.00642	.00649	.00666	.00691	.00740	.00788
11/32	.00618	.00622	.00627	.00636	.00646	.00657	.00681	.00692	.00697	.00704	.00721	.00745	.00794	.00843
3/8	.00672	.00677	.00682	.00691	.00700	.00712	.00735	.00747	.00751	.00758	.00775	.00800	.00849	.00897
13/32	.00727	.00731	.00736	.00745	.00755	.00766	.00790	.00801	.00806	.00813	.00830	.00854	.00903	.00952
7/16	.00781	.00786	.00790	.00800	.00809	.00821	.00844	.00856	.00860	.00867	.00884	.00909	.00958	.01006
15/32	.00835	.00840	.00845	.00854	.00863	.00875	.00898	.00910	.00914	.00922	.00939	.00963	.01012	.01061
1/2	.00890	.00895	.00899	.00909	.00918	.00930	.00953	.00965	.00969	.00976	.00993	.01018	.01067	.01115

1/2	.00890	.00895	.00899	.00909	.00918	.00930	.00953	.00965	.00969	.00976	.00993	.01018	.01067	.01115
17/32	.00944	.00949	.00954	.00963	.00972	.00984	.01007	.01019	.01023	.01031	.01048	.01072	.01121	.01170
9/16	.00999	.01004	.01008	.01018	.01027	.01039	.01062	.01074	.01078	.01085	.01102	.01127	.01175	.01224
19/32	.01053	.01058	.01063	.01072	.01081	.01093	.01116	.01128	.01132	.01140	.01157	.01181	.01230	.01279
5/8	.01108	.01113	.01117	.01127	.01136	.01147	.01171	.01183	.01187	.01194	.01211	.01235	.01284	.01333
21/32	.01162	.01167	.01171	.01181	.01190	.01202	.01225	.01237	.01241	.01249	.01266	.01290	.01339	.01387
11/16	.01217	.01222	.01226	.01235	.01245	.01256	.01280	.01292	.01296	.01303	.01320	.01345	.01393	.01442
23/32	.01271	.01276	.01281	.01290	.01299	.01311	.01334	.01346	.01350	.01357	.01374	.01399	.01448	.01496
3/4	.01326	.01331	.01335	.01344	.01354	.01365	.01389	.01400	.01405	.01412	.01429	.01453	.01502	.01551
25/32	.01380	.01385	.01390	.01400	.01408	.01419	.01443	.01455	.01459	.01466	.01483	.01508	.01557	.01605
13/16	.01435	.01439	.01444	.01453	.01463	.01474	.01498	.01509	.01514	.01521	.01538	.01562	.01611	.01660
27/32	.01489	.01494	.01498	.01508	.01517	.01529	.01552	.01564	.01568	.01575	.01592	.01617	.01666	.01714
7/8	.01544	.01548	.01553	.01562	.01571	.01583	.01607	.01618	.01623	.01630	.01647	.01671	.01720	.01769
29/32	.01598	.01603	.01607	.01617	.01626	.01638	.01661	.01673	.01677	.01684	.01701	.01726	.01775	.01823
15/16	.01653	.01657	.01662	.01671	.01681	.01692	.01716	.01727	.01732	.01739	.01736	.01780	.01829	.01878
31/32	.01707	.01712	.01716	.01726	.01735	.01747	.01770	.01782	.01786	.01793	.01810	.01835	.01883	.01932
1	.01762	.01766	.01771	.01780	.01790	.01801	.01825	.01836	.01841	.01848	.01865	.01889	.01938	.01987

Figure 3-05 is the **'Straight Input'** Excel Spreadsheet. Calculated results are: Bend Allowance, Bend Deduction, Setback, Apex to Inside Radius (Air), ID apex to OD apex, Cut-size, Bend lines from Zero, and the Distances between Bend lines.

Figure 3-05

STRAIGHT:	1.500	5.000	2.000	1.500
BEND ANGLE:	45.000	90.000	115.000	
INSIDE RADIUS:	0.250	0.375	0.500	
MATERIAL THICKNESS:	0.119			
BEND ALLOWANCE:	0.238	0.672	1.109	
BEND DEDUCTION:	0.068	0.316	0.834	
SETBACK:	0.034	0.158	0.417	
'AIR' - APEX TO INSIDE RADIUS:	0.153	0.494	0.972	
ID MATERIAL APEX TO OD APEX:	0.049	0.119	0.187	
CUT SIZE:	12.019			
BEND LINES FROM 0:	1.619	7.074	9.964	12.019
DISTANCES BETWEEN BEND LINES:	1.619	5.455	2.890	2.054

We now have all the information we need to lay out the part full size onto the material from which it will be formed. Let make the part 7" in length shearing a flat blank of material, *Figure 3-06*, from .119" thick steel 7.000" x 12.019".

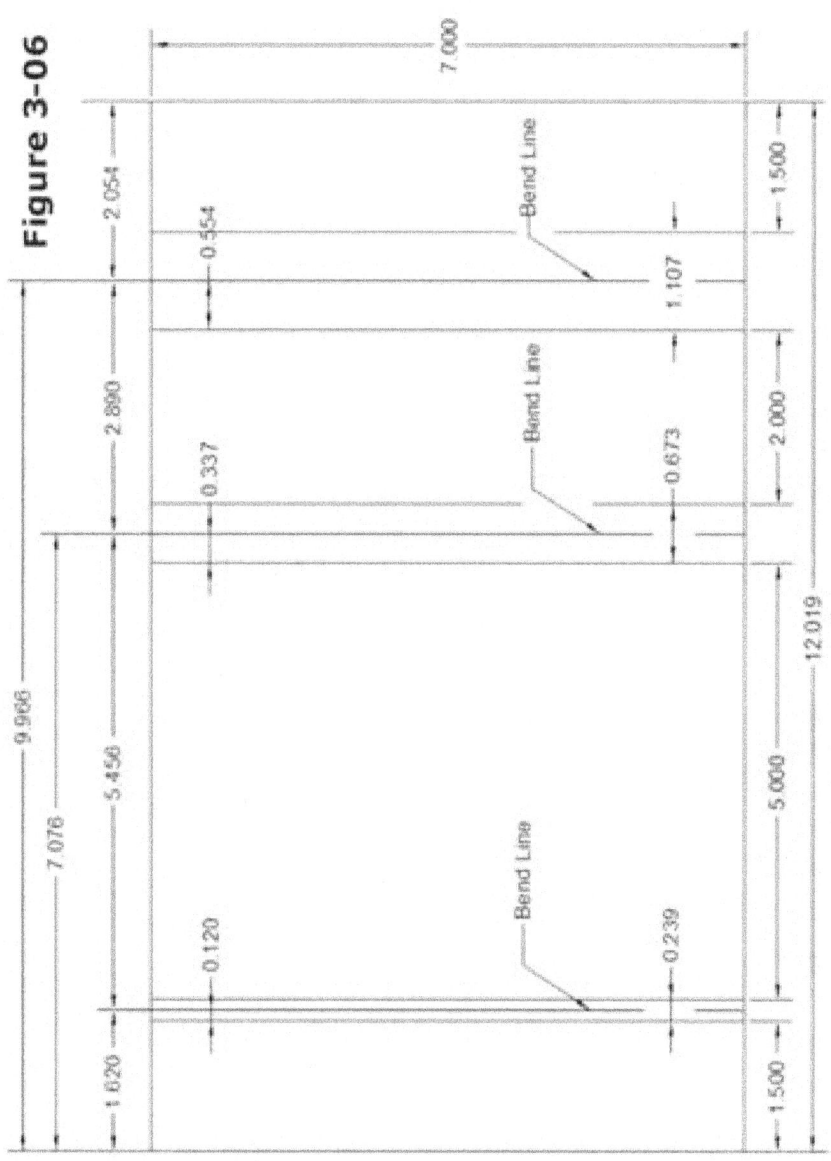

Figure 3-06

The Flat Layout shows:

- ☐ The overall cut size of 12.019" x 7.000".
- ☐ Shown are the three BA calculations of .239", .673" and 1.107".
- ☐ Shown are the four Straight dimensions of 1.500", 5.000", 2.000" and 1.500". Adding all these together we arrive at 12.019".
- ☐ The bend lines (Center of the press-brake tooling) are always located at the center of the 'Bend Allowance' dimension.
- ☐ The center of each bend is dimensioned from the left side of the flat layout are 1.620", 7.076", and 9.966".
- ☐ Alternately the distance between the bend lines are 1.620", 5.456", 2.890", and 2.054" are shown.

Bend Deduction:

One of the components of the 'Bend Deductions' formula is 'Bend Allowance', therefore 'Bend Allowance must be calculated first. 'Bend Deduction' also requires that all dimension be Apex to Apex.
To demonstrate 'Bend Deduction we will be using the same part used in 'Bend Allowance', **Figure 3-07**, only dimensioned apex to apex.

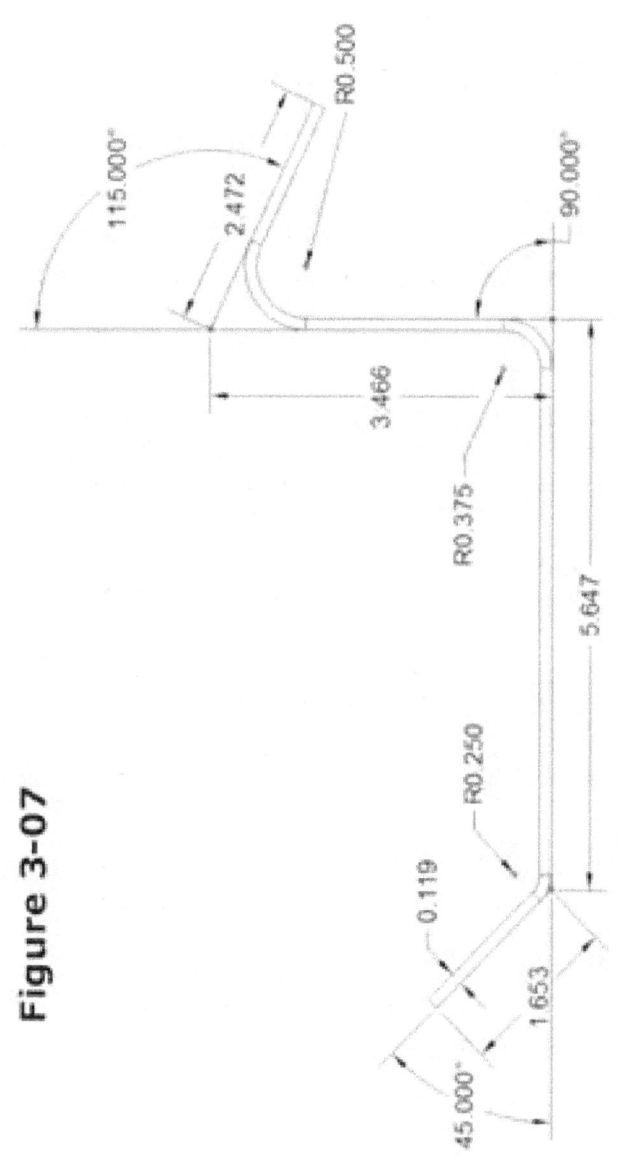

Figure 3-07

Calculate the 'Bend Allowance'

Being the same part, we have previously calculated the 'Bend Allowance' for the following Inside Radius and Angles using .119" thick material.

- ☐ .250" IR - .119" MT – 45-degree angle = .238" BA
- ☐ .375" IR - .119" MT – 90-degree angle = .672" BA
- ☐ .500" IR - .119" MT – 115-degree angle = 1.109 BA

Calculate the 'Bend Deduction'

2 x (tangent (Angle / 2) x (MT + IR)) – BA = BD

- ☐ **2 x (tangent (45 / 2) x (.119" + .250")) – .238" = .068"**
- ☐ **2 x (tangent (90 / 2) x (.119" + .375)) – .672" = .316"**
- ☐ **2 x (tangent (115 / 2) x (.119" + .500")) – 1.109" = .834"**
- ☐

Calculate the 'Cut-size'

Total all the Apex dimension. Total all the 'Bend Deductions. Subtract the total of the 'Bend Deductions from the Total of the Apex dimensions. The result will be the cut size of the part. – **12.019"**

Calculate the 'Setback'

Before calculating the distance between the bend lines. we need to define 'Setback'. 'Setback' is a distance equal to half of the 'Bend Deduction'. 'Setback' places the press-brake punch at the center of a bend when subtracted from its apex dimension.

Bend Deduction / 2   = 'Setback'

- ☐ **.068" / 2   = .034"**
- ☐ **.316" / 2   = .158"**
- ☐ **.834" / 2   = .417"**
- ☐

Calculate the 'Distance between Bend Lines'

To calculate the distance between the bend-lines subtract the setback from each side of a bend.

Apex Dimension – Setback = Distance between Bend Lines

- • **1.653" – (.034")   = 1.620"**
- • **5.648" – (.034" +   = 5.456"**

.158")
- **3.466" – (.158" +**   **= 2.890"**
  .417")
- **2.472" – (.417")**   **= 2.055"**
- **Cut Size**   **= 12.020"**

In our previous flat layout, *Figure 3-06*, using 'Bend Allowance' was heavily dimensioned for instructional purposes. A condensed flat layout, *Figure 3-08*, indication only the distance between the bend lines.

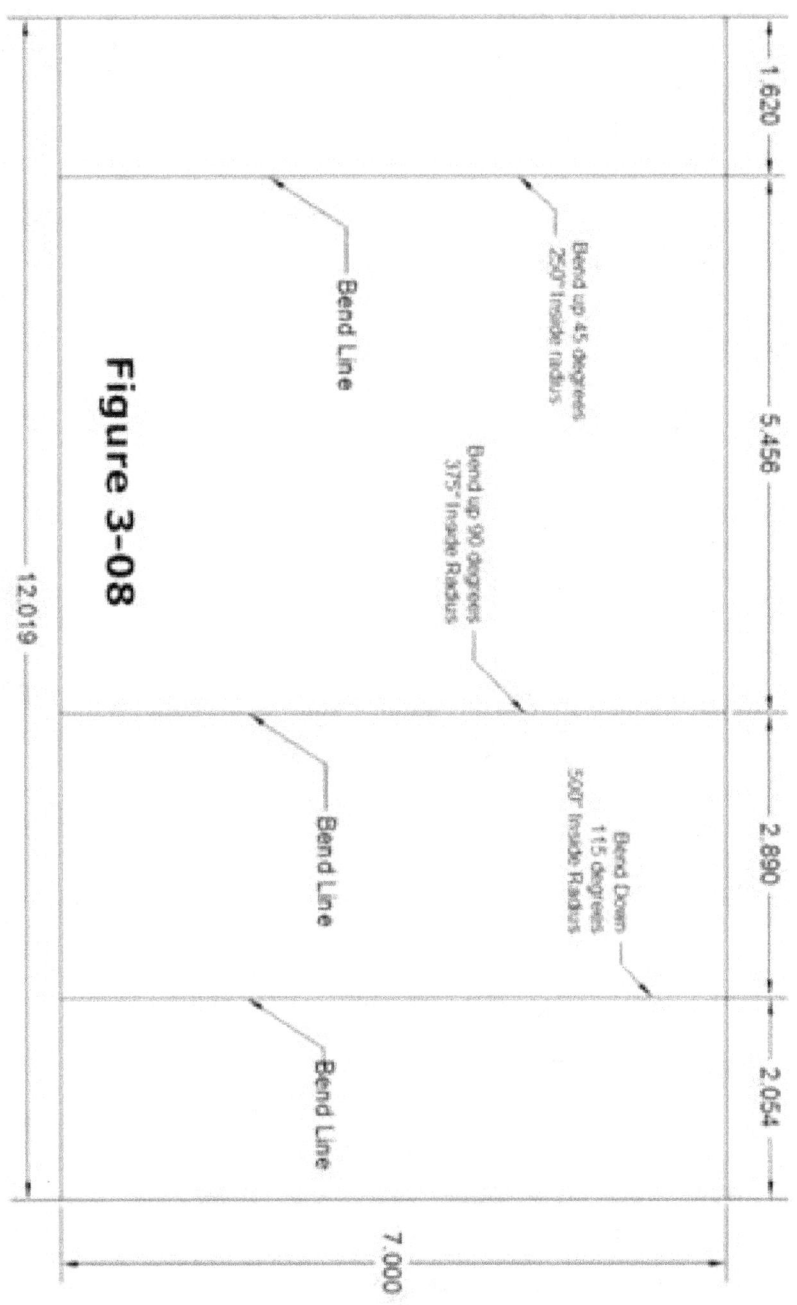

Figure 3-08

Using the 'Apex Input Spreadsheet', **_Figure 3-09_**, simplifies the process by simply entering the material thickness, apex-to-apex dimensions, the bend angles and the inside radii for each bend in the open cells of the spreadsheet.

Figure 3-09

APEX DIMENSION:	1.653	5.647	3.466	2.472
BEND ANGLE:	45.000	90.000	115.000	
INSIDE RADIUS:	0.250	0.375	0.500	
MATERIAL THICKNESS:	0.119			
BEND ALLOWANCE:	0.238	0.672	1.109	
BEND DEDUCTION:	0.068	0.316	0.834	
SETBACK:	0.034	0.158	0.417	
'AIR' - APEX TO INSIDE RADIUS:	0.153	0.494	0.972	
INSIDE MATERIAL APEX TO OD APEX:	0.049	0.119	0.187	
CUT SIZE:	12.020			
BEND LINE FROM 0:	1.619	7.074	9.965	12.020
BETWEEN BEND LINES:	1.619	5.455	2.891	2.055

- 49 -

Bend Allowance Chart

Figure 3-10 is a 'Bend Deduction' calculated from the previous presented formulas. Now you know the Theory behind the chart.

BEND RADIUS	Stock Thickness							
	0.022	0.032	0.040	0.051	0.064	0.091	0.128	0.187
1/32	0.00072	0.00079	0.00086	0.00094	0.00104	0.00125	0.00154	0.00200
1/16	0.00126	0.00135	0.00140	0.00149	0.00159	0.00180	0.00209	0.00255
3/32	0.00180	0.00188	0.00195	0.00203	0.00213	0.00234	0.00263	0.00309
1/8	0.00235	0.00243	0.00249	0.00258	0.00268	0.00289	0.00317	0.00364
5/32	0.00290	0.00297	0.00304	0.00312	0.00322	0.00343	0.00372	0.00418
3/16	0.00344	0.00352	0.00358	0.00367	0.00377	0.00398	0.00426	0.00473
7/32	0.00398	0.00406	0.00412	0.00421	0.00431	0.00452	0.00481	0.00527
1/4	0.00454	0.00461	0.00467	0.00476	0.00486	0.00507	0.00535	0.00582
9/32	0.00507	0.00515	0.00521	0.00530	0.00540	0.00561	0.00590	0.00636
5/16	0.00562	0.00570	0.00576	0.00584	0.00595	0.00616	0.00644	0.00691
11/32	0.00616	0.00624	0.00630	0.00639	0.00649	0.00670	0.00699	0.00745
3/8	0.00671	0.00679	0.00685	0.00693	0.00704	0.00725	0.00753	0.00800
13/32	0.00725	0.00733	0.00739	0.00748	0.00758	0.00779	0.00808	0.00854
7/16	0.00780	0.00787	0.00794	0.00802	0.00812	0.00834	0.00862	0.00908
15/32	0.00834	0.00842	0.00848	0.00857	0.00867	0.00888	0.00917	0.00963
1/2	0.00889	0.00896	0.00903	0.00911	0.00921	0.00943	0.00971	0.01017
17/32	0.00943	0.00951	0.00957	0.00966	0.00976	0.00997	0.01025	0.01072
9/16	0.00998	0.01005	0.01012	0.01020	0.01030	0.01051	0.01080	0.01126
19/32	0.01051	0.01058	0.01065	0.01073	0.01083	0.01105	0.01133	0.01179
5/8	0.01107	0.01114	0.01121	0.01129	0.01139	0.01160	0.01189	0.01235
21/32	0.01161	0.01170	0.01175	0.01183	0.01193	0.01214	0.01245	0.01289
11/16	0.01216	0.01223	0.01230	0.01238	0.01248	0.01268	0.01298	0.01344
23/32	0.01269	0.01276	0.01283	0.01291	0.01301	0.01322	0.01351	0.01397
3/4	0.01324	0.01332	0.01338	0.01347	0.01357	0.01378	0.01407	0.01453

Bend Allowance per One Degree

Test Bending:

Test Bending is the most accurate way to determine 'Bend Deduction' where high tolerance is required or when you are deviating from the 'Air Bend' range, bending exotic materials, using small radii in thick materials, or simply want to catalog you own log of results.

The same part, **Figure 3-11**, used to describe 'Bend Allowance' and 'Bend Deduction' is used to detail 'Test Bending'. The middle 90-degree bend with the .375" Inside Radius is used to go through the process.

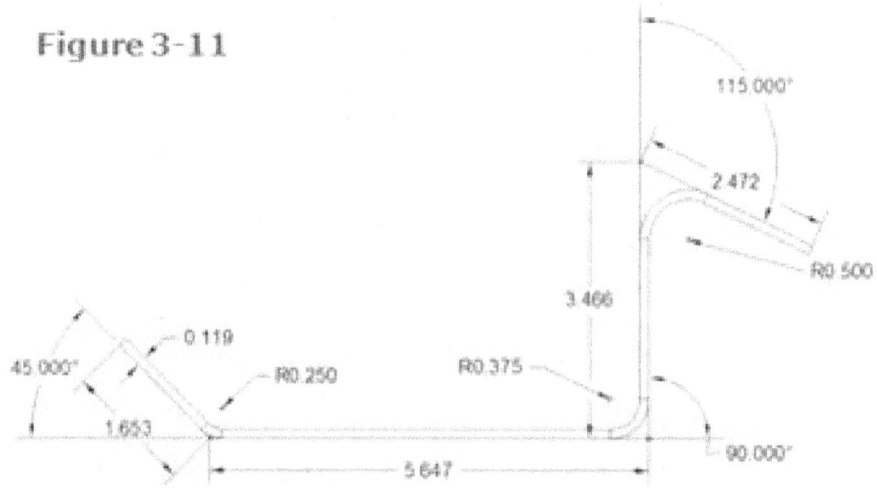

Figure 3-11

Create a 'Test Bend Record' for each 'Test Bend' for future usage. There a seven (7) Known parameters and three calculated results:

- ☐ **Grade of the material**   **Steel – Hot Rolled**
- ☐ **Material thickness**   **.119"**
- ☐ **Bend angle**   **90 degrees**
- ☐ **Inside radius**   **.375"**
- ☐ **Test Blank size**   **Suitable x Suitable**
- ☐ **Description of Top Punch**   **0.750" Diameter**
- ☐ **Description of Bottom Die**   **1.000" Open vee die**
- ☐
- ☐ **First Flange Apex Dimension**   _____
- ☐ **Second Flange Apex Dimension**   _____
- ☐ **Bend Deduction**   _____

Test Bend Procedure:

Using the Information in the 'Test Bend Record'.

☐ In this test example cut a test blank 6.000" x any convenient
length from the chosen material and thickness.
☐ Set up the press-brake with a 0.750" round top punch and a 1.000" open vee bottom die.
☐ Setup the press-brake to form the material to a 90-degree angle in one (1) cycle. **Note:** More than one cycle of the Press Brake will taint the results.
☐ Bend the test piece to the above requirements.
  o   The test bend location can be formed anywhere along its length. Near center is recommended.
  o   The test blank does however need to be square in relationship to the press-brake.
☐ After bending, measure each flange from the Outside apex to the edge of the material.
  o   The result is: **3.937" and 2.379".**
  o   Adding the previous results in: **6.316".**
  o   Subtract the original blank cut size of **6.000"** from
     6.316"
  o   The result is: **0.316"** - The 'Bend Deduction'.

The process was repeated for the 45 and 115-degree bends:

☐ 45 degree - .250" Inside Radius = .068"   BD
☐ 115 degree - .500" Inside Radius = .834" BD

Setback = BD/2

☐ 45 degree - .250" Inside Radius = .034"   SB
☐ 115 degree - .500" Inside Radius = .417" SB

Figure 3-12 illustrates the 'Test Bending' Excel spreadsheet. Using the Excel spread result in additional information, most pertinent pieces of information are the Bend Allowance and Setback.

TEST BENDING

DISCRIPTION OF MATERIAL: Steel - Hot Rolled

DISCRIPTION OF TOP PUNCH   .750" Round
DISCRIPTION OF BOTTOM DIE: 1.000" Vee

1.571
BEND ANGLE   90.000
MATERIAL THICKNESS: 0.119
INSIDE RADIUS: 0.375

TEST LENGTH: 6.000
FLANGE ONE - APEX DIMENSION: 3.937
FLANGE TWO - APEX DIMENSION: 2.379

BEND DEDUCTION: 0.316
SETBACK: 0.158

BEND ALLOWANCE: 0.672

'AIR' - APEX TO INSIDE RADIUS: 0.494
INSIDE MATERIAL APEX TO OD APEX: 0.119

'Bend Deduction' Excel Spreadsheet

The 'Bend Deduction' spreadsheet, *Figure 3-13*, simplifies the math process by calculating the cut size, location of the bend lines, and the setback.

There are only two inputs that need be entered into the 'Bend Deduction' spreadsheet. They are the outside apex to outside apex dimensions and the results of a 'Test Bend'.

There is no need to enter the Bend angle, Inside Radius, or Material thickness as they are properties of the 'Test Bending' process which determined the Bend Deduction.

The apex dimensions entered here are for the same part used throughout. The Bend Deductions entered was theoretically from a 'Test Bend' scenario.

Note that the calculated results are the same no matter which spreadsheet, 'Straight Input', 'Apex Input', or 'Bend Deduction Input' was applied.

"OD" APEX 2 APEX:	1.653	5.647	3.466	2.472
BEND DEDUCTION:	0.068	0.316	0.834	
SETBACK:	0.034	0.158	0.417	
CUT SIZE:	12.020			
BEND LINE FROM 0:	1.619	7.074	9.965	12.020
BETWEEN BEND LINES:	1.619	5.455	2.891	2.055

Figure 3-13

Notes

Part IV

Deck Openings

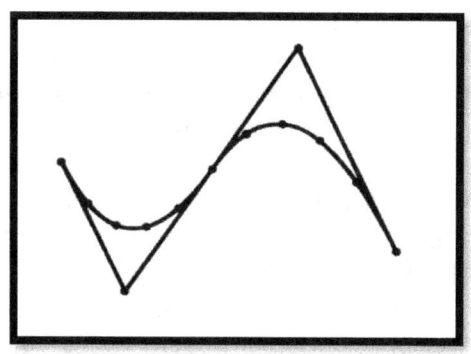

Architectural drawings routinely represent the size and location of features such as hatches and their coaming by way of a single drawn line. For example, the 'Construction Plan and Profile', **Figure 5-01,** for the Bezier 35 shows only the size, location, and height of the coaming above the deck.

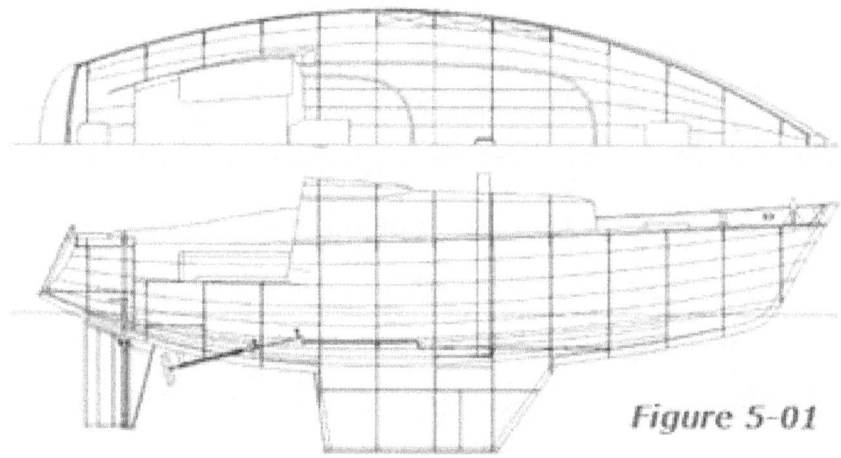

Figure 5-01

Detailed are two types of deck Coamings. They are not the only solutions, but give the reader a reasonable starting point to create architectural pleasing deck openings.

- A Raised Coaming Deck Hatch.
  - With Solid Cover
  - With Skylight Cover
- Production Hatch Coaming for mounting Purchased Hatches.

Raised Coamings:

The 'Shop Drawings', *Figure 5-02,* outlines the forward deck hatch assembly of the Bezier 35.

Starting with the coaming, notice the crown which follows the curve of the deck transversely. The coaming widest point is 4.344" to 4.146" at the centerline, these dimensions were taken from the transverse frame at the hatch location. The hatch coaming will be 0.187" thick and is fabricated in two pieces with the seams on the hinge sides.

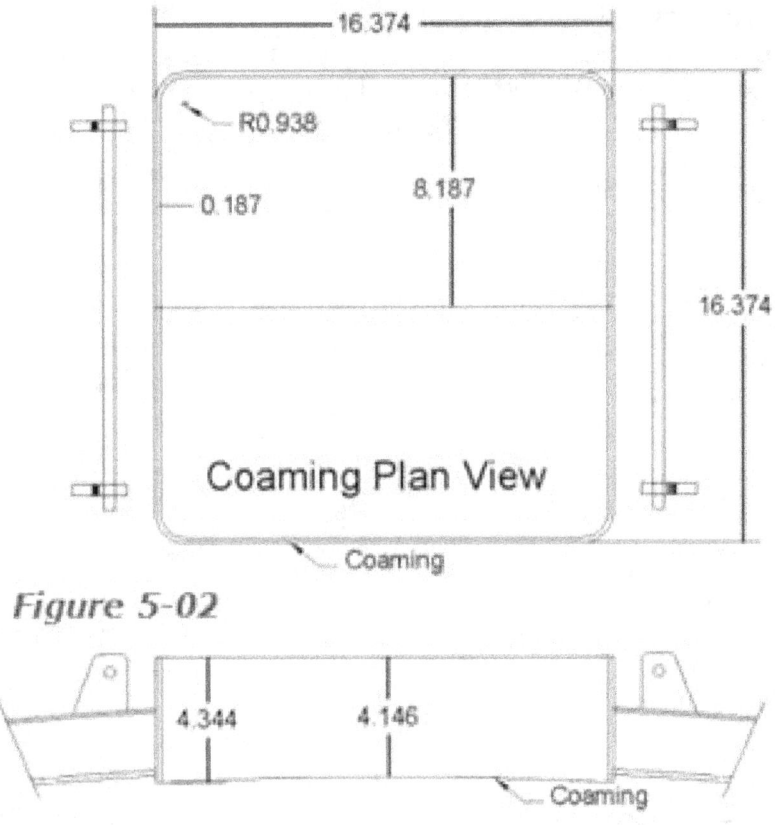

Figure 5-02

From these drawings all the dimensional information necessary to layout the coaming can be entered into the development spreadsheet, *Figure 5-03,* to calculate the cut size and bend line locations. The apex dimensions are: 8.187" x 16.375" x 8.187". The bend angles are both 90 degrees and the inside radius of the bend is 0.9375". Making the diameter of the Top punch 1.875" in diameter.

Figure 5-03

APEX DIMENSION:	8.187	16.375	8.187
BEND ANGLE:	90.000	90.000	
INSIDE RADIUS:	0.937	0.937	
MATERIAL THICKNESS:	0.187	0.187	
BEND ALLOWANCE:	1.601	1.601	
BEND DEDUCTION:	0.647	0.647	
SETBACK:	0.323	0.323	
'AIR' - APEX TO INSIDE RADIUS:	1.124	1.124	
INSIDE MATERIAL APEX TO OD APEX:	0.187	0.187	
CUT SIZE:	31.455	31.455	
BEND LINES FROM 0:	7.864	23.592	31.455
BETWEEN BEND LINES:	7.864	15.728	7.864

From the results calculated a flat layout, **Figure 5-04**, can be hand drawn. It will provide all the information necessary for a Metal Fabricator to form the Forward hatch coaming for The Bezier 35.

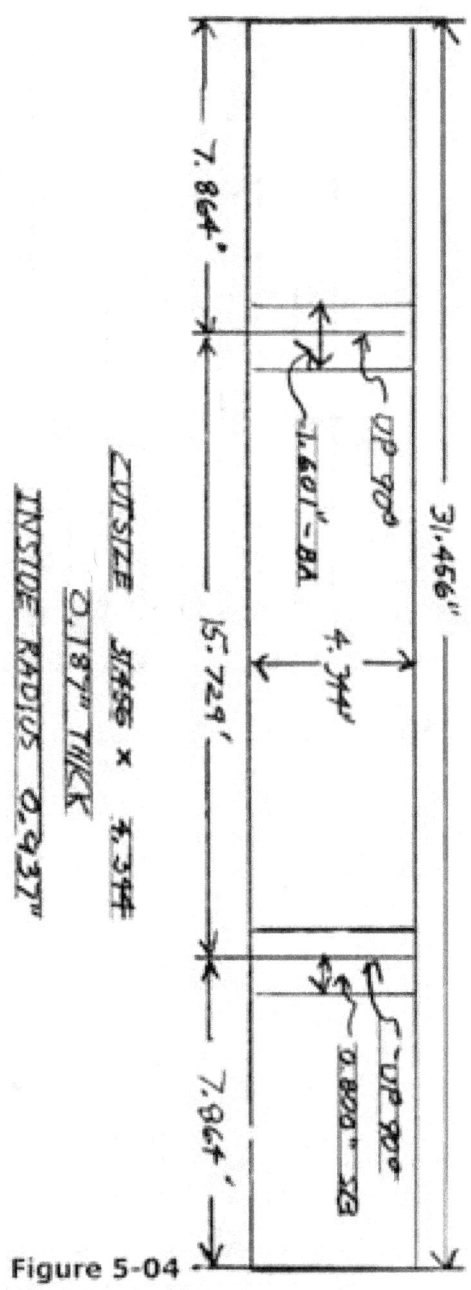

Figure 5-04

Onto this Raised Coaming two (2) Hatch Covers styles will be detailed.

- A solid cover fabricated from 11 ga Sheet Material.
- A Port Light cover fabricated from Angle Iron.

Solid Hatch Cover

The 'Shop Drawing' in **Figure 5-05** shows the finished outside dimensions of the hatch cover at 17.624" x 17.624" square. Its flange depth is 1.500" and material thickness is 0.119" – 11ga. The radius dimension used for the corners was chosen to coincide with the section dimensions of 3.000" Schedule 40 steel pipe, which in turn determined

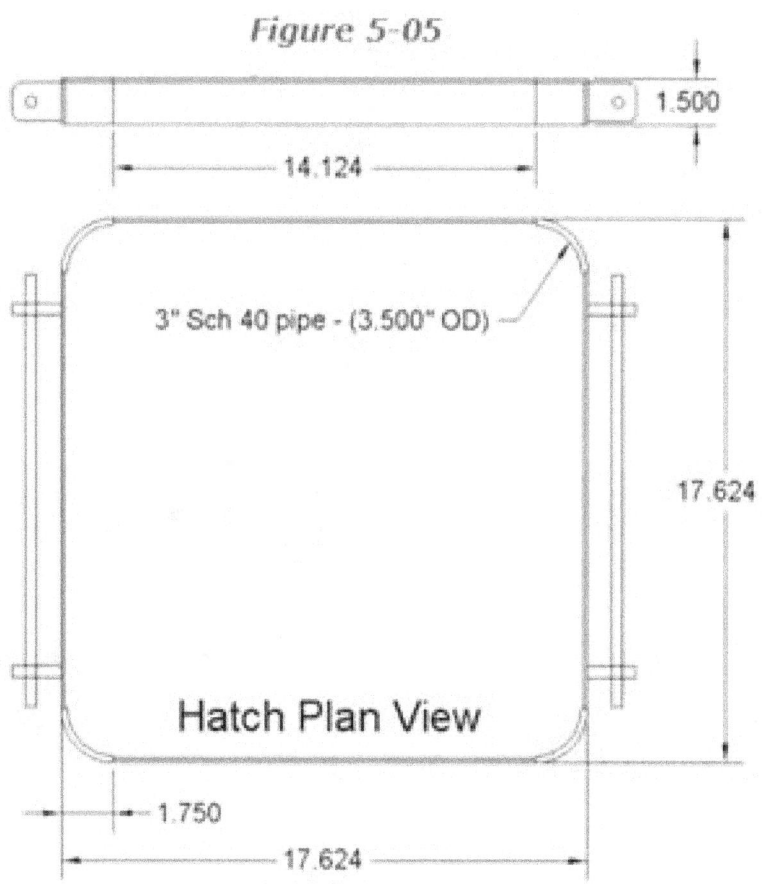

Figure 5-05

the radius of the coaming, previously calculated. Quarter sections of this pipe are used in a secondary welding operation to complete the flange corners.

This 'Shop Drawing' provides us with all the information necessary to enter into the various spread-sheets to calculate 'Inside Radius', Cut-Size, bend lines, and notch depths.

For this hatch cover I chose to use a one-inch bottom die to soften the touch and appearance of the covers bent-on flanges. A one-inch open vee bottom die used to form 0.119" thick steel produces an 'Inside Radius' of 0.156". *Figure 5-06*

Figure 5-06

The next spreadsheet, **Figure 5-07,** calculates the cut size. Since the Hatch cover is square only one calculation is required.

Figure 5 - 07

APEX DIMENSION:	1.500	17.625	1.500
BEND ANGLE:	90.000	90.000	
INSIDE RADIUS:	0.156	0.156	
MATERIAL THICKNESS:	0.119		
BEND ALLOWANCE:	0.328	0.328	
BEND DEDUCTION:	0.222	0.222	
SETBACK:	0.111	0.111	
'AIR' - APEX TO INSIDE RADIUS:	0.275	0.275	
INSIDE MATERIAL APEX TO OD APEX:	0.119	0.119	
CUT SIZE:	20.182		20.182
BEND LINES FROM 0:	1.389	18.792	20.182 / 1.389
BETWEEN BEND LINES:	1.389	17.403	1.389

- 63 -

This part needs to be corner notched, **Figure 5-08,** to allow the flanges to be bent. After entering the material thickness, bend deduction, and the OD apex dimension of the flange into its spreadsheet the notch depth is calculated at 1.397".

- The Material Thickness is 0.119" as previous.
- The Bend Deduction is found in the previous Spreadsheet.
- The Flange Dimension is 1.500" as before.

Figure 5 - 08

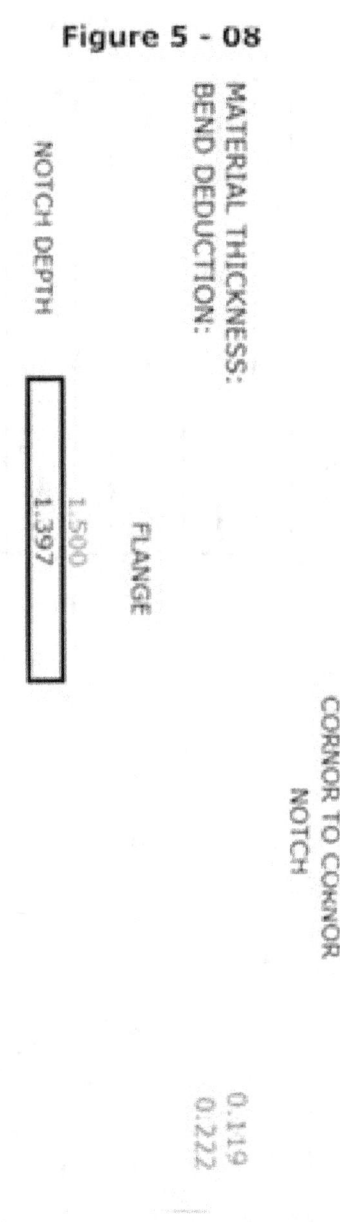

With the dimensional information contained in the 'Shop Drawing' and the spread-sheet calculations completed a hand sketch can be drawn, **Figure 5-09**, to complete the layout process.

Start by drawing a square 20.180" x 20.180". To this add the notch information. Refer back to, **Figure 5-05,** *and* note the dimension 14.124". This is the straight distance between the starts of the radii.

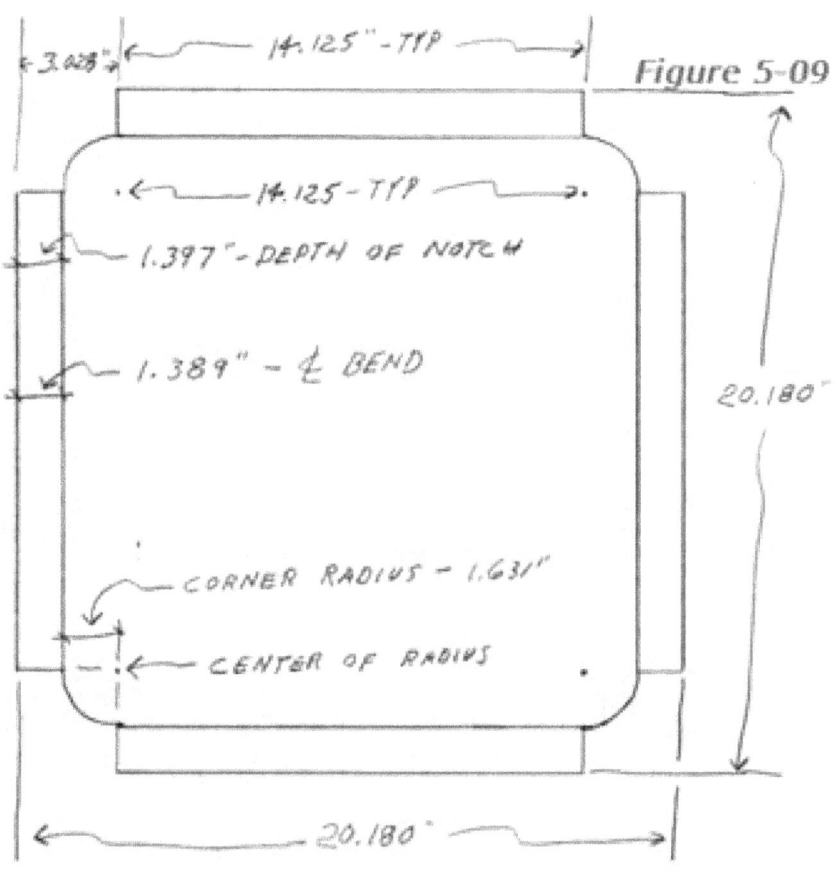

Figure 5-09

Subtracting this number (14.125") from the cut-size of the part (20.180") dividing this result by two. The product is 3.028". This is the notch depth in way of the corner radius.

The notch depth at the flange has previously been calculated at 1.397".

- 65 -

The last notch layout factor is to drawn in the radius at each corner. The center of the radius is determined by extending the 'Start of the Radius' lines at each corner. Their crossing point will be the radius center. The radius dimension drawn will be the distance between the fore mentioned radius center to the notch depth. To calculate this distance, subtract 1.397", the notch depth of the flange, from 3.028", the notch depth to the start of the radius. The resulting radius to be drawn is 1.631". Finally note the bend line dimensions is calculated at 1.389".

Skylight Hatch Cover

The outside dimensions and the radius corner dimensions of the Skylight hatch cover, **Figure 5-10**, is exactly the same as the solid hatch cover. The difference lies in the fabrication. The perimeter of this type of hatch cover will be fabricated from 1.500" x 1.500" x 0.125" angle iron instead of sheet material. Using an angle iron frame for the perimeter of the hatch allows clear polycarbonate to be attached allowing light into the interior.

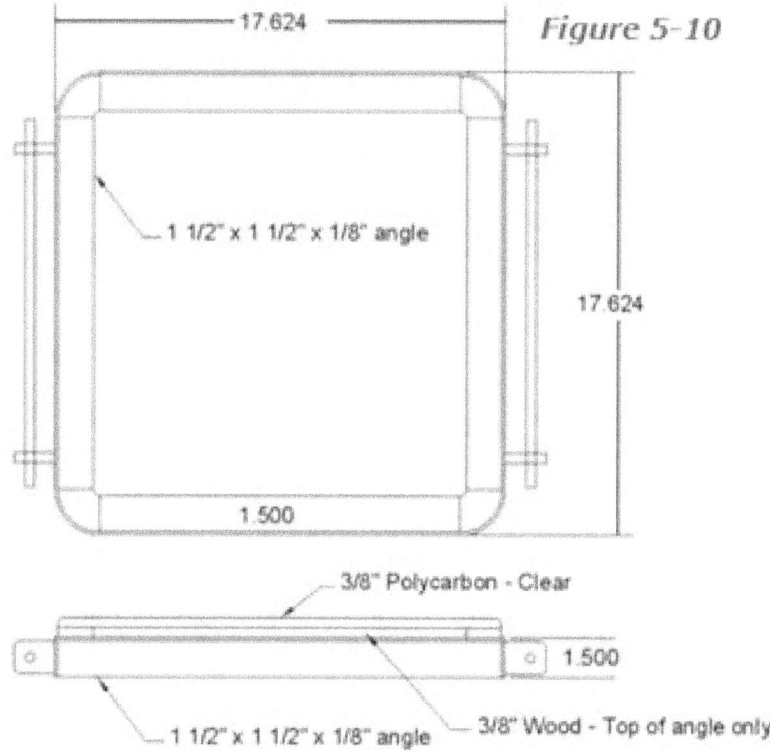

Figure 5-10

The outside radius dimension of this hatch corner is also the same as the solid hatch at 1.750". Making the 'Inside Radius' used in the spread-sheet calculation 1.625". The hatch cover will be formed from a single piece of angle iron with a single weld seam in the center of one of its sides. The spread-sheet calculating the cut size and bend lines locations are shown in **Figure 5-11.**

APEX DIMENSION:	8.812	17.625	17.625	17.625	8.812
BEND ANGLE:	90.000	90.000	90.000	90.000	
INSIDE RADIUS:	1.625	1.625	1.625	1.625	
MATERIAL THICKNESS:	0.125	0.125			
BEND ALLOWANCE:	2.637	2.637	2.637	2.637	
BEND DEDUCTION:	0.863	0.863	0.863	0.863	
SETBACK:	0.432	0.432	0.432	0.432	
'AIR' - APEX TO INSIDE RADIUS:	1.750	1.750	1.750	1.750	
INSIDE MATERIAL APEX TO OD APEX:	0.125	0.125	0.125	0.125	
CUT SIZE:	67.047	25.142	41.904	58.666	67.047
BEND LINES FROM 0:	8.380	16.762	16.762	16.762	8.380
BETWEEN BEND LINES:	8.380	16.762	16.762	16.762	

Figure 5 - 11

The flat layout, **Figure 5-12**, can be marked directly onto the 1.500" x 1.500" x 0.125" angle iron from which the frame is fabricated. A 'Bend Allowance' of 2.637" is the amount of material to be removed from one of the flanges to facilitate forming at each corner. Forming the angle iron frame can be accomplished by hand forming around a section of 3" pipe.

Figure 5 - 12

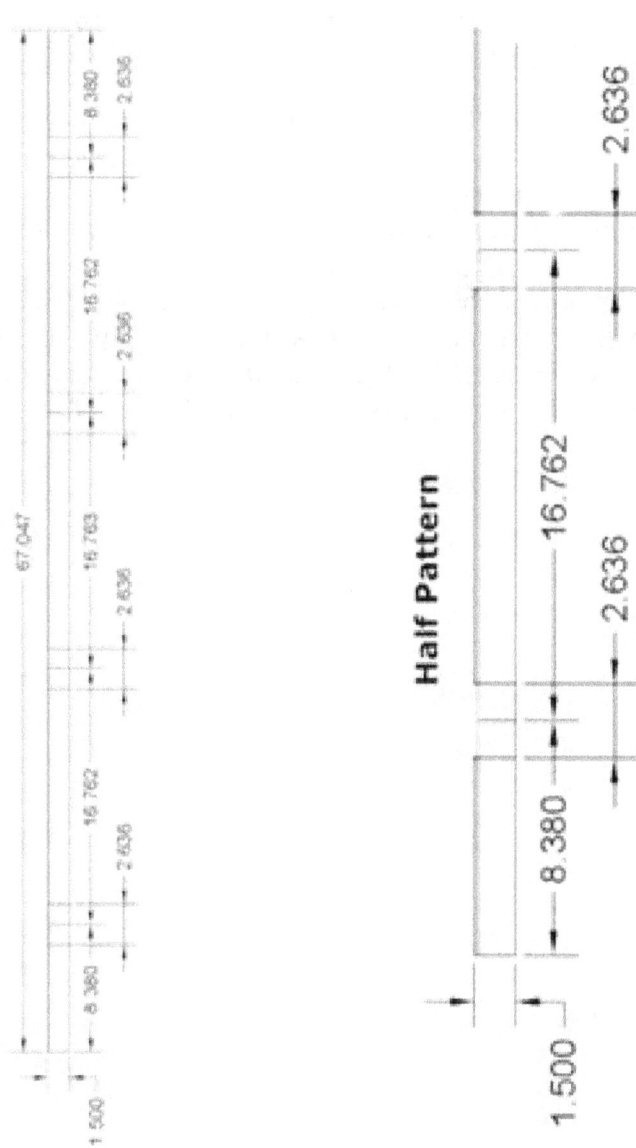

Production Hatch's:

Fabricated steel hatches keep the character of a metal builds, but this alone should not preclude the installation of commercially available products.

The solution illustrated here joins a Production Deck Hatch, designed for fiberglass construction, to a steel hull is a fabricated custom **'J-Style'** framework into the boats structure - ***Figure 5-13.*** There are certain principal features that I considered important from an architectural point of view to incorporate into the design.

- A **'Revel'** of 1/4" between the perimeter of the Production hatch and the perimeter of the **'J-Style'** coaming is visually necessary.
- The height of the coaming above the deck would need to be held to a practicable minimum.
- The exterior of the assembly must flow into the interior of the cabin ceiling finish in a pleasing way. To this end **note** the wooden trim placed in the pocket of the **'J-Style'** section of the coaming as see from below – Inside the hull.

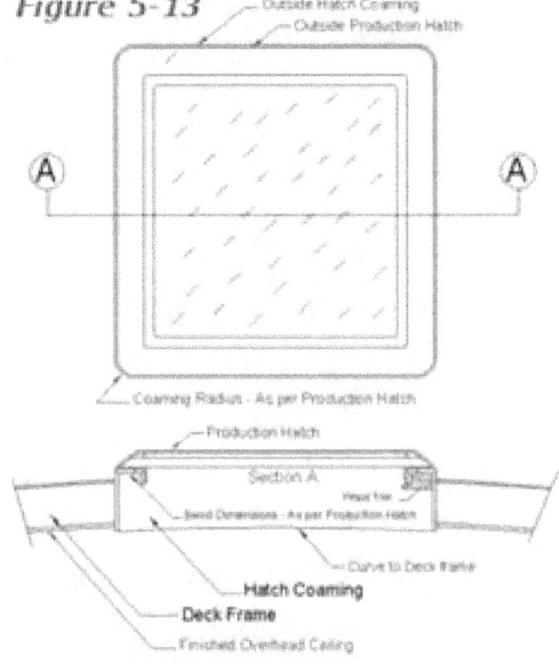

Figure 5-13

With the above criteria in mind the dimensions of the chosen Production Hatch are:

- Overall dimensions are 20.500" x 20.500".
- Outside corner radius are 2.500"
- Deck opening is 18.000" x 18.000"
- Deck opening radius is 1.250"

The 'Shop Drawing' has two views - **_Figure 5-14._** The top view shows the relationship between the Production Hatch and the custom fabricated **'J-Style'** coaming and how it flows pleasingly into the

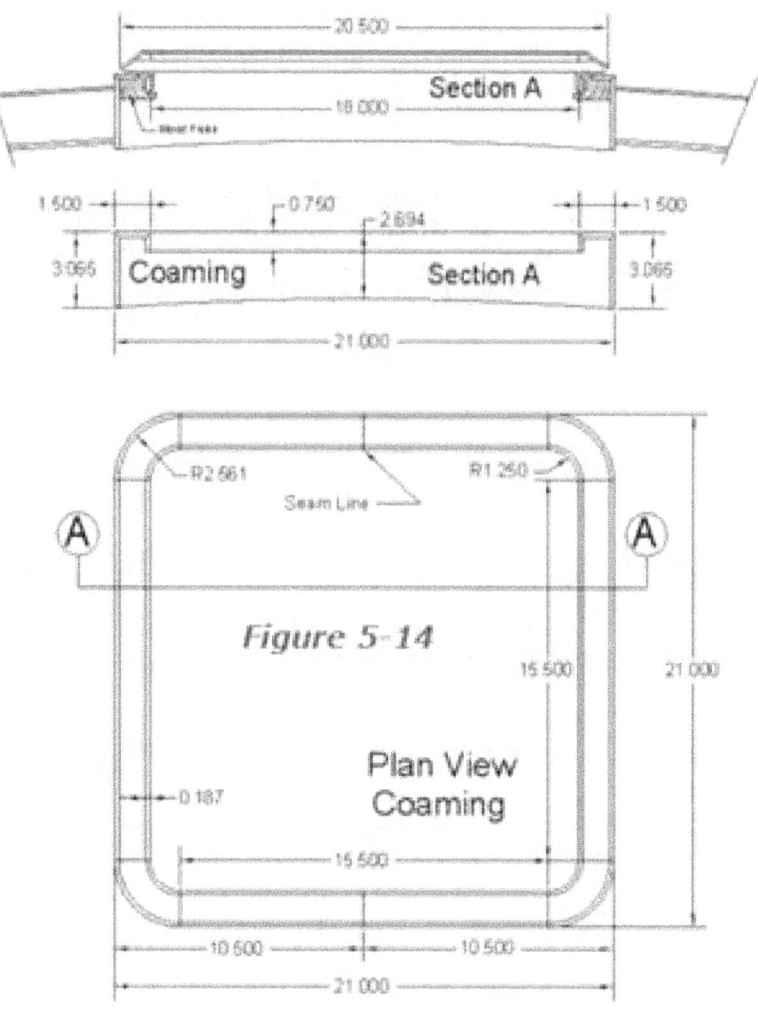

Figure 5-14

Plan View
Coaming

Interior finish. **Note** also how the **'J-Style'** coaming matches the curvature of the deck.

The middle section separates the framework from all the other features of the top drawing. The **'J- Style** fabricated coaming is dimension at 0.750" x 1.500" x 3.065". The distance between the apex dimensions of the coaming is 21.000" which leaves a **'Revel'** of 0.250" around the perimeter of the 'Production Hatch' at 20.500". The depth of the coaming at the centerline of the beck is 2.694".

The bottom sketch is the plan view of the **'J- Style'** coaming. A seam line is show since the coaming frame will be fabricated in two sections. Also shown is the 'Inside Radius' of the 'Production Hatch' (1.250"), which is one of the dimensions necessary for alignment between the 'Production Hatch' and the custom **'J-Style'** coaming. Also called out is the 'Inside Radius' (2.561") for the outside of the coaming.

To develop the flat layout, first calculate the Inside Radius of 0.187" thick steel, using a vee bottom die opening of 1.000". The spread-sheet, **Figure 5-15**, calculates the Inside Radius at 0.156". **Note** that the ratio between a vee bottom die of 1.000" and 0.187" thick material is 5.3, the beginning of the Bottom bending mode.

INSIDE RADIUS CALCULATIONS
FOR
AIR - BOTTOM BENDING RANGE

Figure 5-15

MATERIAL THICKNESS:	0.187
BOTTOM DIE OPENING:	1.000
BOTTOM DIE OPENING AT 8 X MATERIAL THICKNESS:	1.496
BOTTOM DIE OPENING AT 7 X MATERIAL THICKNESS:	1.309
BOTTOM DIE OPENING AT 6 X MATERIAL THICKNESS:	1.122
BOTTOM DIE OPENING AT 5 X MATERIAL THICKNESS:	0.935

AIR TO BOTTOM BENDING RANGES

INSIDE RADIUS:	0.156
BOTTOM DIE OPENING VS MATERIAL THICKNESS:	5.3
BOTTOM BENDING PERCENTAGE INDICATOR:	83%

We now have all the dimensions we need to enter into the 'Apex Input' spreadsheet to calculate the cut size and to locate the bend line to form the **'J-Style'** coaming section.

To layout the coaming we need to calculate the cut size for both axis of the part. Starting with the apex dimension that will form the **'J-Style'** channel enter (0.750" – 1.500" – 3.065") as per the 'Shop Drawing'. Both bends are 90 degrees. Using an 'Inside Radius' of 0.156". The results are calculated in **Figure 5-16**.

Figure 5 - 16

	0.750	1.500	3.065
APEX DIMENSION:	0.750	1.500	3.065
BEND ANGLE:	90.000	90.000	
INSIDE RADIUS:	0.156	0.156	
MATERIAL THICKNESS:	0.188		
BEND ALLOWANCE:	0.377	0.377	
BEND DEDUCTION:	0.311	0.311	
SETBACK:	0.156	0.156	
'AIR' - APEX TO INSIDE RADIUS:	0.344	0.344	
INSIDE MATERIAL APEX TO OD APEX:	0.188	0.188	
CUT SIZE:	4.693		
BEND LINES FROM 0:	0.594	1.783	4.693
BETWEEN BEND LINES:	0.594	1.189	2.909

Next calculated is the cut size for the radius bends along the other axis **Figure 5-17** of the part. Going back to the 'Shop Drawing' we find the apex dimensions to enter are 10.500" x 21.000: x 10.500". Both bends are 90 degrees with and 'Inside Radius' of 2.561".

APEX DIMENSION:	10.500	21.000	10.500
BEND ANGLE:	90.000	90.000	
INSIDE RADIUS:	2.561	2.561	
MATERIAL THICKNESS:	0.188	0.188	
BEND ALLOWANCE:	4.149	4.149	
BEND DEDUCTION:	1.348	1.348	
SETBACK:	0.674	0.674	
'AIR' - APEX TO INSIDE RADIUS:	2.749	2.749	
INSIDE MATERIAL APEX TO OD APEX:	0.188	0.188	
CUT SIZE:	39.303		
BEND LINES FROM 0:	9.826	29.478	39.303
BETWEEN BEND LINES:	9.826	19.652	9.826

Figure 5 -17

The cut size of the **'J-Style'** framework onto which the 'Production Hatch will mount is 4.693" x 39.303".

The next calculations will be the width and depth of the cutouts which relieve the corners for forming. The width of the notch is already calculated and is logically the 'Bend Allowance' of 4.149".

The depth of that notch is calculated in, **Figure 5-18**, since this notch goes into the second bend whose apex dimension is 1.500".

Figure 5 - 18

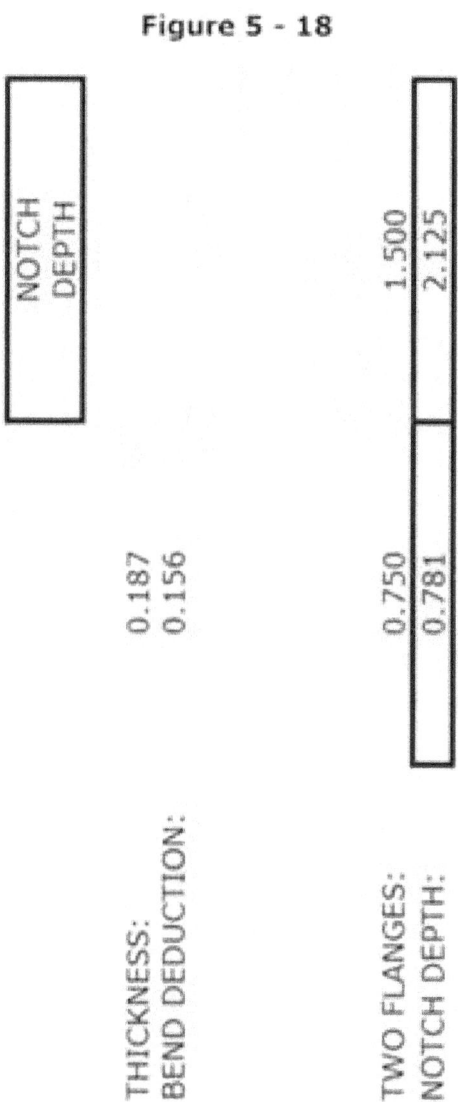

3

We now have all the information necessary to represent the flat layout, **Figure 5-19**, directly on the material from which it will be formed if we so chose.

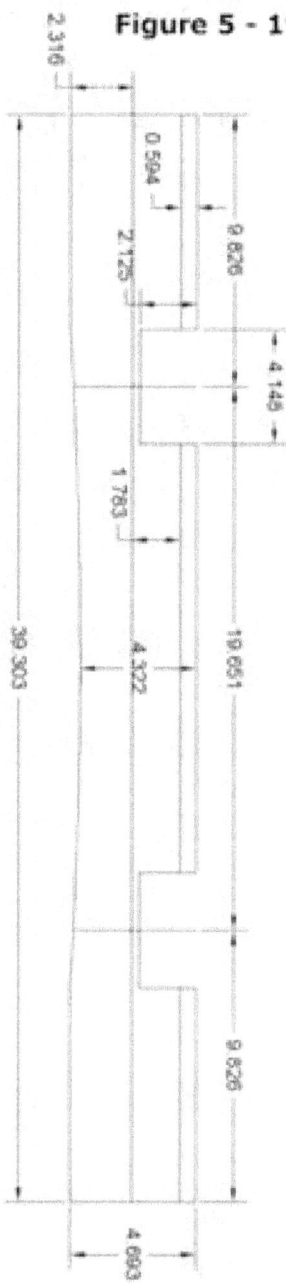

Figure 5 - 19

The last calculation, **Figure 5-20**, is the press brake tonnage to form the part. In this spreadsheet we enter the material thickness, bottom die opening, material factor of steel at (1), and the length of the part at 39.000".

The calculated results indicate that the bends pressures is in the 'Bottom' bending range at 5.3 times the material thickness. Tonnage required to bend .187" thick material, 90 degrees, using a 1.000" bottom die opening, is 60.802 tons.

Figure 5 - 20

Notes:

Part VI

Port-Lights

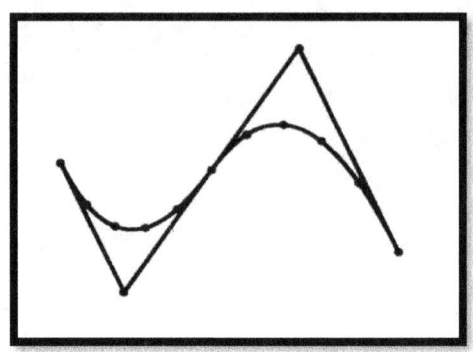

I once read an online forum post by a steel boat advocate. Their view was "That a steel boat is what it is – Just let it all hang out there". In the context of the Post, I interrupted their meaning to be that steel boats cannot be as beautiful and gracious as boats constructed from other building materials. "So, just let it all hang out there".

I say, "It is all in the details". Up to this point I have not detailed half-hearted or quick and easy fabrications. As I move on to window ports, I will not be detailing simple window port solutions such as thru-bolting polycarbonate sheet directly to the cabin shell plating or welding an angle to the shell plating than attaching a sheet of polycarbonate to it. I will be detailing comprehensive 'Shop Drawings' for window port that are both Architecturally pleasing on the exterior while flowing seamlessly into the interior finish.

The interior profile drawing, **Figure 6-01**, locates and sizes the window port for the 'Bezier 35'.

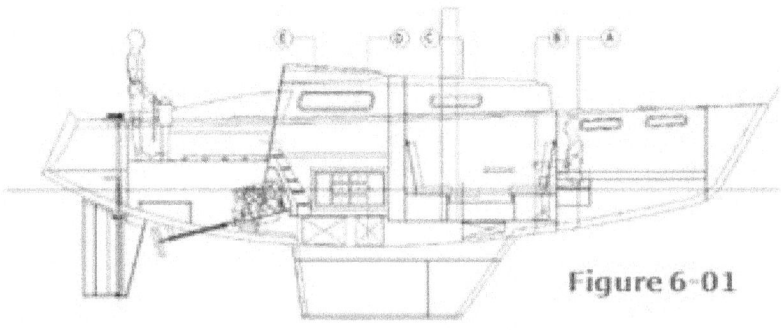

Figure 6-01

Fixed Coaming Port-Light:

Figure 6 - 02

Beginning with a 'Shop Drawing' of the aft house Port-Light, **Figure 6-02**, as viewed from the outside. The close parallel lines represent a 3/16" thick steel coaming. Small circles represent thru bolt holes in the mounting frame for the polycarbonate window.

As always, the section views, **Figure 6-03**, detail the bulk of the information. The left section indicates an extended or overhanging coaming past the shell plating, while the right shows the coaming ground flush to the shell plating.

After studying the section drawings, it can be seen how this design connects the exterior watertight requirements on the outside of the hull with the interior finish requirements on the inside of the hull.

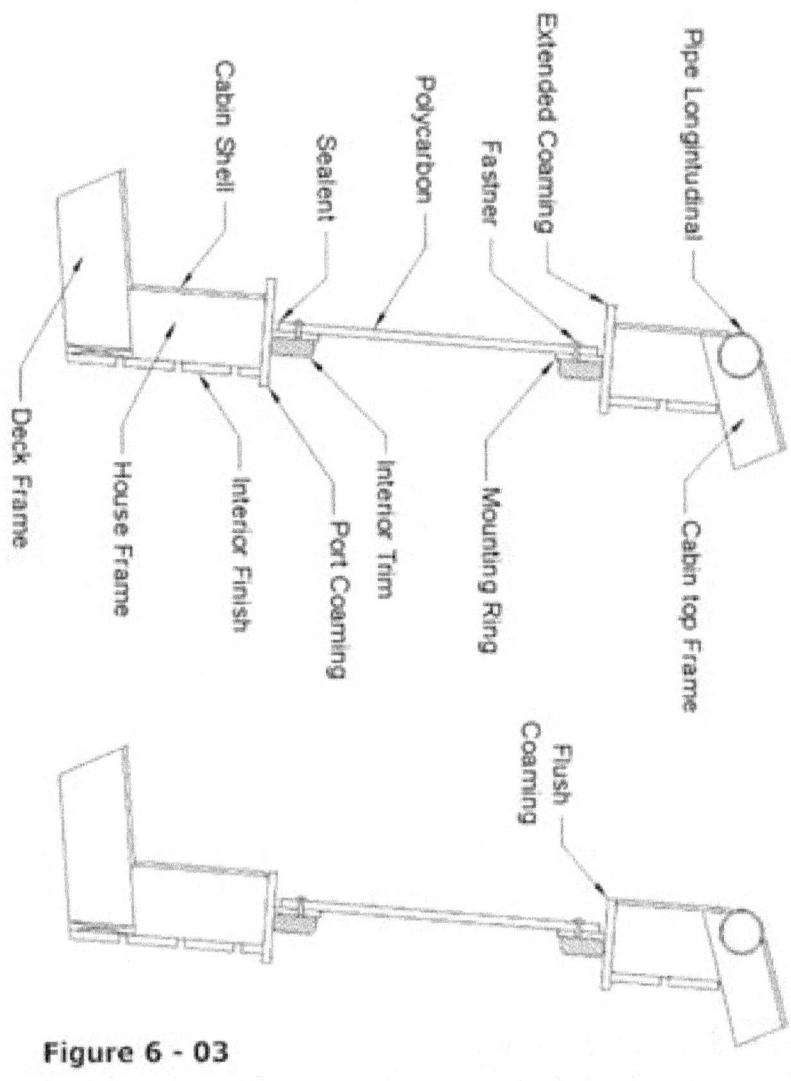

Figure 6 - 03

The coaming:

Dimensional information in, *Figure 6-04*, is a representation of your full size 'Shop Drawing' derived from the Architectural Drawings.

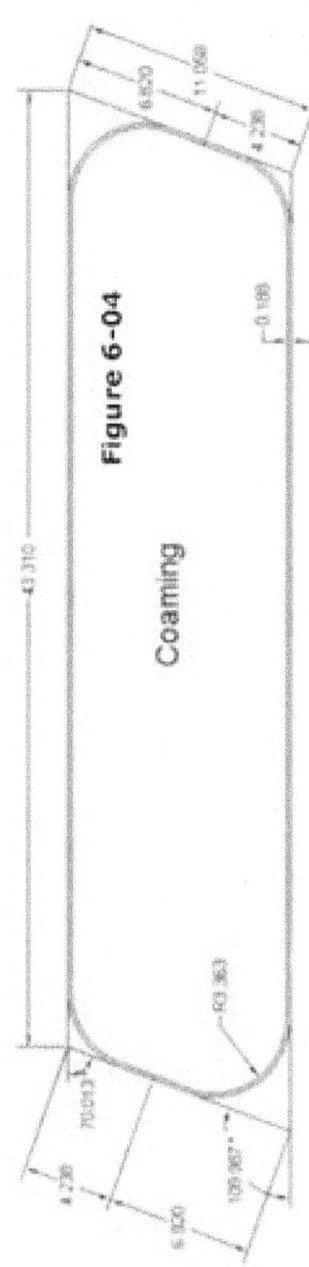

Figure 6-04

Coaming

- This coaming will be fabricated in two half sections from 3/16"
  thick steel.
- The seams are located on the center of the short straight sides.
- The apex dimensions are 4.238" at the obtuse angle, 6.820", at the acute angle, and 43.310" between.
- The inside radius is 3.375" for all four corners. The acute angle is 70.013 degrees while the obtuse angle is 109.987 degrees.
- Spreadsheet calculation are shown in *Figure 6-05*

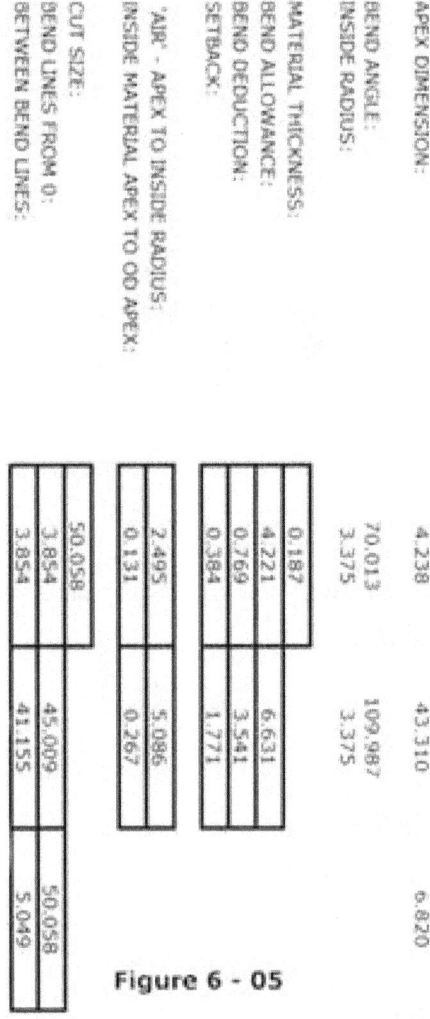

Figure 6 - 05

The hand sketch, **Figure 6-06**, documents the flat layout, providing all the necessary dimensions to cut and form the coaming for this fixed port light. **Note:** 'Bump Forming' is used to form the relatively large radii specified.

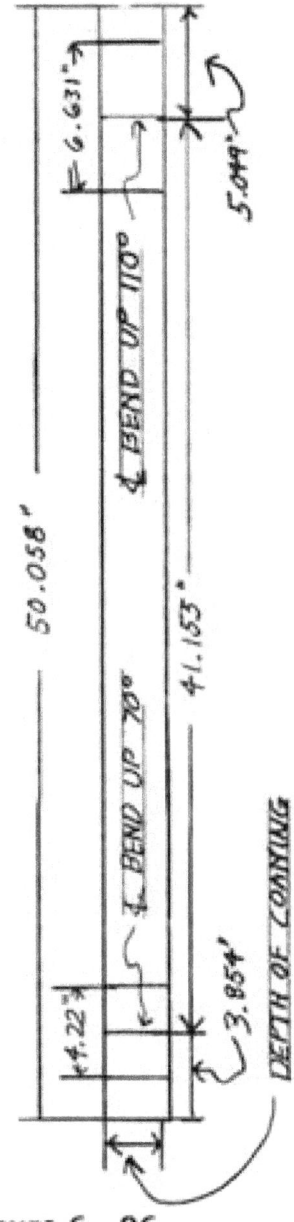

Figure 6 - 06

Mounting Flange:

The mounting flange has already been developed via your 'Shop Drawing'. All you need to do is transfer your full size 'Shop Drawing' to the material from which it will be cut. 'Cnc' cutting of this part would be ideal. Here is one way to cut the mounting flange by hands.

- Referring to, **Figure 6-07**, top drawing, cut a rectangular blank
  46.554" x 10.017".
- Next shears 3.643" off the corners at upper left and lower right, the bottom drawing.
- See **Figure 6-08** for final Band Saw cutting.

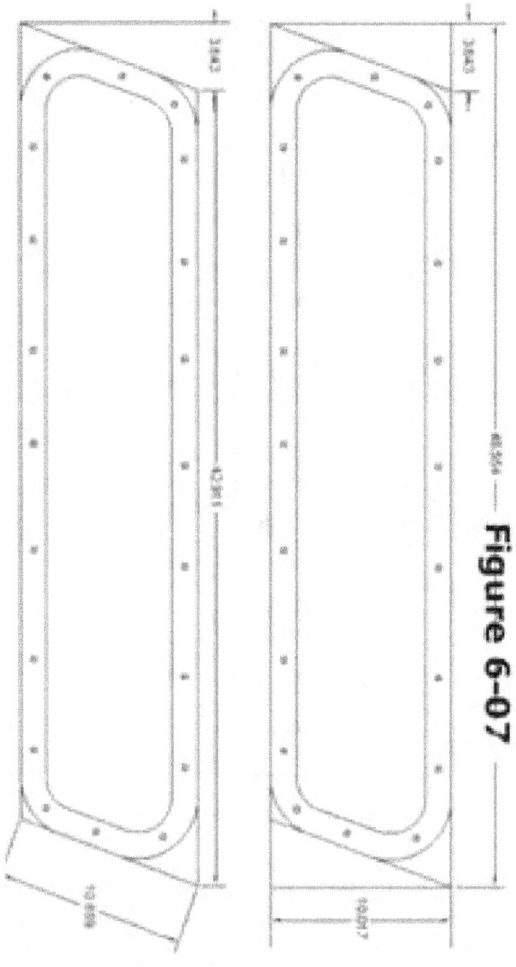

Figure 6-07

Note: The mounting flange is fully welded to the coaming on the inside on the hull only.

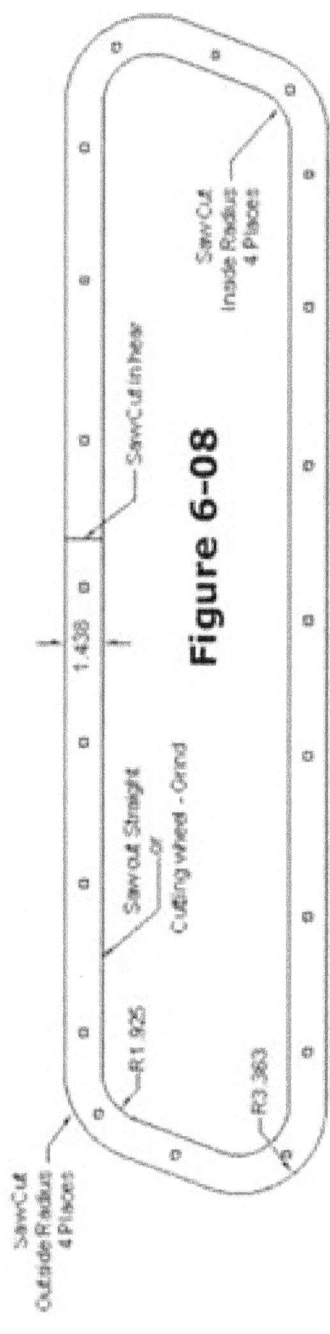

Figure 6-08

Fixed Mitered Port-Light

The Fixed Mitered Window Port, **Figure 6-09**, is a hard corner design that adds styling to the port by designing in wooden trim to accent the ports border

Figure 6-09

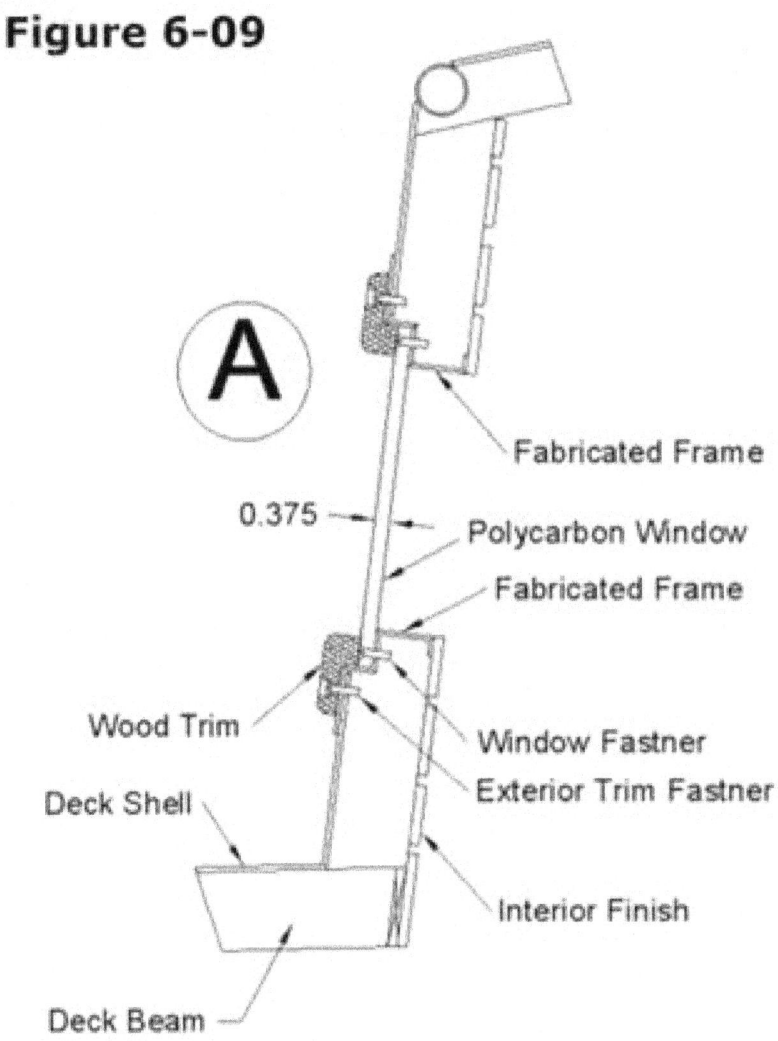

The fabricated coaming, **Figure 6-10**, shown in profile is constructed from four (4) mitered fabricated framing channels using the apex dimensions of 42.911" x 12.920". Radius corners (1.750") shown are rounded off after the four-sided frame is fabricated.

Figure 6-10

Frame Section Detail

12 ga Port Frame

The section view of the coaming frames are detailed in **Figure 6-11.** **Note:** Two of the dimensions are dimensioned 'Inside material thickness to Outside of the material. We will need to add the material thickness to these dimensions to determine the outside apex dimension before entry into the 'Apex' dimension spreadsheet.

Figure 6-11

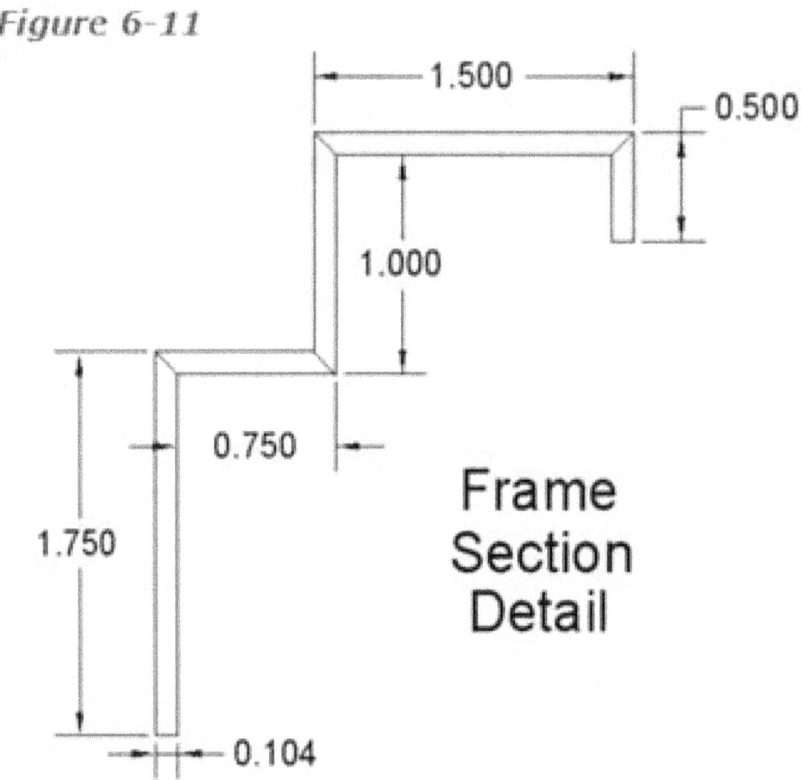

Frame Section Detail

The 'Inside Radius' of the bends will be based on a vee die opening of 0.750". While the air-bottom bend ratio will be determined by the material thickness of 0.104".

Entering this information in, **Figure 6-12**, we find that the 'Inside Radius' for the bend is 0.117". Bottom die opening vs material thickness is 7.2, placing this bend in the air bending mode.

Figure 6-12

INSIDE RADIUS CALCULATIONS
FOR
AIR – BOTTOM BENDING RANGE

MATERIAL THICKNESS: 0.104
BOTTOM DIE OPENING: 0.750

BOTTOM DIE OPENING AT 8 X MATERIAL THICKNESS: 0.832
BOTTOM DIE OPENING AT 7 X MATERIAL THICKNESS: 0.728
BOTTOM DIE OPENING AT 6 X MATERIAL THICKNESS: 0.624
BOTTOM DIE OPENING AT 5 X MATERIAL THICKNESS: 0.520

AIR TO BOTTOM BENDING RANGES

INSIDE RADIUS: 0.117

BOTTOM DIE OPENING VS MATERIAL THICKNESS: 7.2
BOTTOM BENDING PERCENTAGE INDICATOR: 113%

The press-brake tonnage calculation, **Figure 6-13**, estimates the tonnage required to form a 90-degree bend in 0.104" thick material using a 0.750" open vee bottom die at 30.8 tons for a bend 48" long.

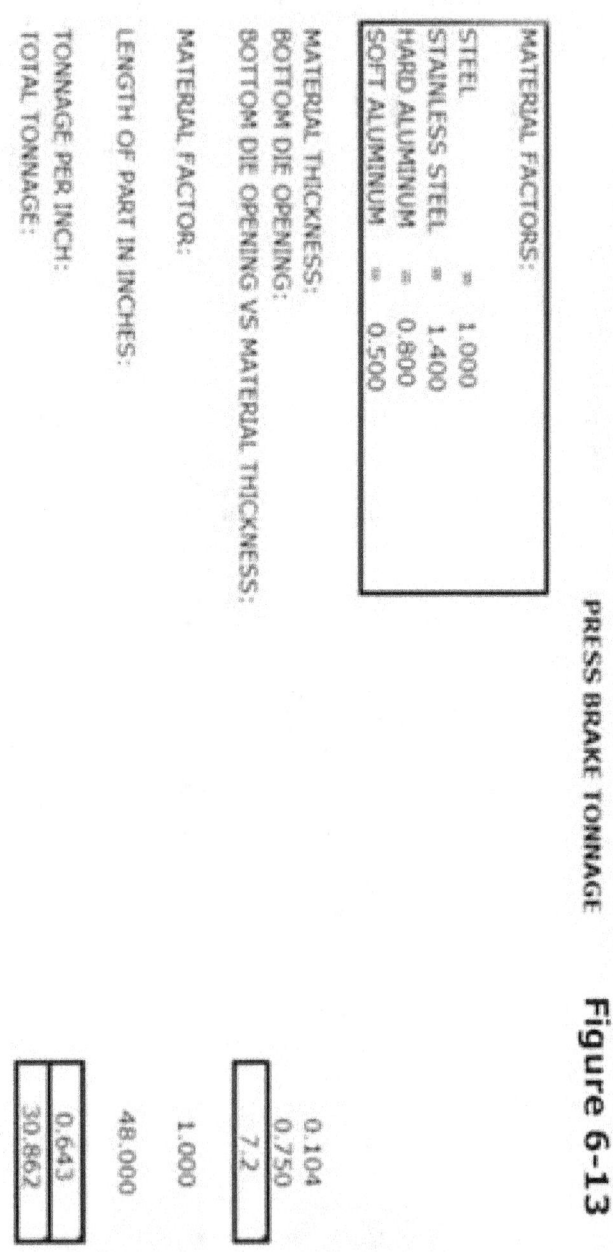

PRESS BRAKE TONNAGE   **Figure 6-13**

MATERIAL FACTORS:

STEEL	=	1.000
STAINLESS STEEL	=	1.400
HARD ALUMINUM	=	0.800
SOFT ALUMINUM	=	0.500

MATERIAL THICKNESS:	0.104
BOTTOM DIE OPENING:	0.750
BOTTOM DIE OPENING VS MATERIAL THICKNESS:	7.2
MATERIAL FACTOR:	1.000
LENGTH OF PART IN INCHES:	48.000
TONNAGE PER INCH:	0.643
TOTAL TONNAGE:	30.862

We now have all the information to enter into the 'Apex to Apex' Excel spreadsheet, **Figure 6-14**, to calculate the cut size.

Figure 6-14

	0.500	1.500	1.104	0.654	1.750
APEX DIMENSION:	0.500	1.500	1.104	0.654	1.750
BEND ANGLE:	90.000	90.000	90.000	90.000	
INSIDE RADIUS:	0.117	0.117	0.117	0.117	
MATERIAL THICKNESS:	0.104				
BEND ALLOWANCE:	0.257	0.257	0.257	0.257	
BEND DEDUCTION:	0.185	0.185	0.185	0.185	
SETBACK:	0.093	0.093	0.093	0.093	
'ABX' - APEX TO INSIDE RADIUS:	0.221	0.221	0.221	0.221	
INSIDE MATERIAL APEX TO OD APEX:	0.104	0.104	0.104	0.104	
CUT SIZE:	4.966				4.966
BEND LINES FROM 0:	0.407	1.722	2.640	3.309	
BETWEEN BEND LINES:	0.407	1.315	0.919	0.669	1.657

- 115 -

Bend-Form the Window frames as per the following.

- The apex dimensions on the 'Shop Drawing' **Figure 6-10** are 42.911" x 12.920".
- Add at least 2.000" to the length of the preceding dimensions making the length of the parts approximately 45.000" and 15.000" long.
- The developed width of the Window Frame is 4.966".
- Therefore, the cut size for the upper and lower frame will be 45.000" x 4.966". The cut size for the side of the frame will be 15.000" x 4.966".
- Bend-form the frame sections as per the 'Shop Drawings' and calculations.
- After Bend-Forming, use a 'Cut-Off Saw to miter the ends of each frame section to the given angle in the 'Shop Drawings', while holding the Apex dimension of 42.911" x 12.911".
- Weld the Frame section together, and round of the ends of the frame as per 'Shop Drawing'.

Installing Production Ports:

Production windows port are intended for installation into fiberglass and wood builds. There is a minimum and maximum range for installation based on the thickness of the shell. The working range of production window ports hover between 0.500" to something over 1.000" depending on the manufacture.

Steel hulls have a framing system. The width of the framework in all practically would be no smaller than 1.500" in depth, well beyond the maximum adjustment for Production Hatches. *Figure 6-15* details a resolution.

The key here is the 'Port Backer'. The 'Port Back' can be any number of different materials that would be available in thickness that would accommodate Production Port-lights. Wood and colored Poly-carbon come immediately to mind.

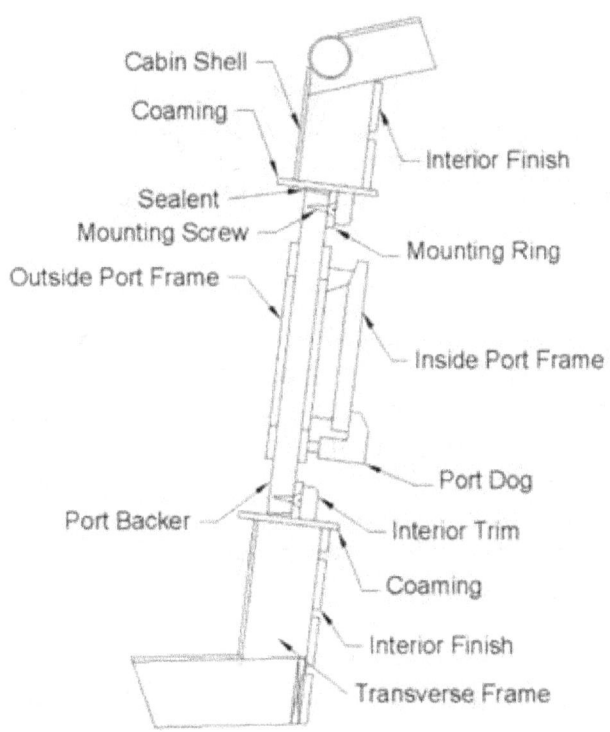

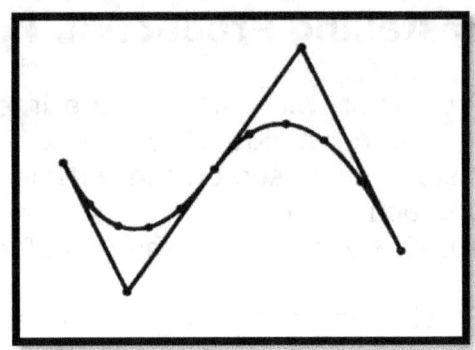

Summary:

Metal fabrication is a trade. It is a lot like being a carpenter. Instead of working with wood you are working with metal. In both trades there are clear principles and procedures of how wood and metal products are designed, fabricated, and assembled. I wanted you, the 'one of' metal boat builder, to realize that you do not have to 'Tuck and Fit' your components, but accurately developed them the first time, every time.

The fabrications used in the book are certainly not the only configurations possible nor are they the simplest. The solutions represented here take time and effort and are in keeping with the architecture of Bezier true round steel designs. They serve as examples upon which you can expand or simplify. Your imagination in this respect is limitless.

The book is subtitled **'Sheetmetal Pattern Development'** to appeal to the majority of Metal Fabricators who are not metal boat builders. To these fabricators this book takes the mystery out of Precision pattern development thru a comprehensive understanding of 'Bend Allowance and 'Bend Deduction. In short, a course in Precision Sheet-metal Layout. If you are working in a sheet metal shop, plan to, or just want to perfect your layout skills **'Applied Metal Boatbuilding Methods'** will assist you in that endeavor.

Excel spreadsheets used in this book are available free of charge and are 'Instantly' downloaded to you from:

metalsailboats.com

Other Books

Hard Copies at Amazon

True Round Metal Boat Building

True Round Metal Boat Design

Steel Mast Design and Fabrication

Notes

www.ingramcontent.com/pod-product-compliance
Lightning Source LLC
Chambersburg PA
CBHW081739220526
45468CB00008B/2160